SCORPIONS
and
VENOMOUS
INSECTS
of the
SOUTHWEST

by

Erik D. Stoops & Jeffrey L. Martin

D1113759

GOLDEN
WEST ☼
PUBLISHERS

Front cover illustration of a Giant Desert Hairy Scorpion and text illustrations by Jeffrey L. Martin

Library of Congress Cataloging-in-Publication Data
Stoops, Erik D.,
    Scorpions and Venomous Insects of the Southwest / by Erik Stoops and Jeffrey Martin.
        p. cm.
    Includes Bibliographical References, Glossary and Index.
    1. Arthropoda—Southwestern States. 2. Arthropoda, Poisonous—Southwestern States. I. Martin, Jeffrey L. II. Title.
    III. Title: Scorpions and Venomous Insects of the Southwest.
QL434.52.S68S76   1995   95-1918
595—dc20   CIP

Printed in the United States of America

2nd Printing © 1997

ISBN #0-914846-87-6

Golden West Publishers, Inc.
4113 N. Longview Ave.
Phoenix, AZ 85014, USA
(602) 265-4392

Table of Contents

Full Color Insert is located between pages 64 and 65

Acknowledgments 5
Introduction 6

ARACHNIDS 7
  Scorpions 8
    Bark Scorpion 10
    Devil Scorpion or Striped-tail Scorpion 12
    Giant Desert Hairy Scorpion 14
  Scorpion-like Arachnids 15
    Pseudoscorpions & Chernetids 15
    Windscorpion, Sunspider 17
    Whipscorpion, Vinegaroon 18
    Tailless Whipscorpion 21
  Spiders 23
    Tarantula 26
    Wolf Spider 27
    Black Widow Spider 29
    Brown or Violin Spider 31
    Crab Spider 33
  Ticks and Mites 35
    Adobe Tick, Fowl Tick 36
    Rocky Mountain Wood Tick 37
    Brown Dog Tick 38
    Scabies Mite 39
    Chigger Mite, Red Bug 40

MYRIAPODS
  Centipedes 41
    House Centipede 42
    Giant Desert Centipede 44
  Millipedes 46
    Desert Millipede 47

(Table of Contents continued on next page . . .)

(Continued from previous page)

INSECTS 48

Bees, Wasps & Ants...51
Leafcutter Bee 52
Sonoran Desert Bumble Bee 54
Carpenter Bee 55
Digger Bee...56
Honey Bee 57
Africanized Honey Bee (Killer Bee) 59
Velvet Ant 60
Bald-faced Hornet...62
Yellowjacket 63
Umbrella Wasp or Paper Wasp 65
Cicada Killer 67
Pepsis Wasp, Tarantula Hawk 68
Mud Dauber 70
Harvester Ant 71
Southern Fire Ant 72
Field Ant, Red Ant 73

Beetles 75
Blister Beetle 76
Darkling Beetle 77
Bombardier Beetle 78

True Bugs & Flies 80
Conenose Bug 80
Giant Water Bug 82
Mosquito 84
Horse Fly and Deer Fly 87

Fleas & Lice 89
Human Flea 90
Head Louse and Body Louse 92
Crab Louse 93

Moths 95
Puss Moth Caterpillar 95
Buck Moth 97

Symptoms & First Aid for Bites and Stings 99
Index 101
Glossary 105
Reading List..107
About the Authors 109

Acknowledgments

Ms. Laura Baartz, Entomological Society of America. Ms. Marilyn Bloom of Arizona State University Microbiology Lab. Dave Mills of the State Agricultural Department. Carl Hayden Bee Research Center. Dr. Gerald M. Loper. Carl Olson of the University of Arizona. Howard Lawler and the Staff at the Arizona Sonoran Desert Museum. Norman Grim, Ph.D. of Northern Arizona State University. Donald Kunkel, M.D. of Good Samaritan Regional Poison Control Center. Arizona National Parks and Monuments. Francis Brady, Gretchen Stohl. J.W. Stewart, J.K. Olson, and Garland McIlveen at Texas A&M University. Our friends and Family. Terry Christopher. Special thanks to Tom C. Cook for research assistance. First Aid information was provided by the Good Samaritan Regional Medical Center — Samaritan Regional Poison Center, Phoenix, Arizona.

Additional acknowledgment for manuscript review is given to Steve Prchal, Director of Sonoran Arthropod Studies, Inc., and Roy A. Barnes, M.S., Dept. of Biology, Scottsdale Community College.

Introduction

The animals described in this guide belong to the phylum **Arthropoda,** the largest in the animal kingdom. They are an ancient and highly successful group due to their diversity, adaptability and relatively small size. They can endure extremes in temperature, and survive prolonged periods of food deprivation. Tough exoskeletons protect them from dehydration and physical damage. Multiple jointed appendages have evolved into antennae, pincers, legs, paddles, and wings. The arthropods have exploited all inhabitable niches, being the first animals to leave the sea, as well as the first to take to the air.

Southwestern arthropods in particular are diverse in variety and habit because of the many different habitats present in the region. From the mountains to the desert floor and the shoreline, all of the land-dwelling arthropod classes are represented. Arachnids, myriapods (millipedes, centipedes, etc.,) and insects can be found throughout the Southwestern United States and Mexico.

The purpose of this guide is to introduce the reader to some of the venomous and poisonous arthropods that occur in the Southwest, in a portable, user friendly format. We will describe and illustrate the species which are likely to be encountered by hikers, campers and residents in this region. We will also discuss parasitic and disease carrying species and those arthropods which appear venomous, but in actuality are not.

The reader may find it helpful to review the Glossary on pages 105 & 106.

We have included first-aid information in the text and in a reference section of its own: (*See Symptoms and First Aid for Bites and Stings, page 99*).

To those who are interested in collecting or keeping arthropods, we stress that familiarizing oneself through extensive reading and consulting experts in the field is the best way to stay out of physical danger (*See Reading List, page 107*).

Some of these animals are quite dangerous; we do not recommend handling them.

Arachnids

Class Arachnida

This class of arthropods derives its name from the beautiful weaving girl, Arachne, of Greek mythology whose tapestries rivaled those of the Gods. The jealous goddess Athena turned her into a spider; and so this story explains how spiders came to spin their beautiful and ornate webs.

Class Arachnida is the largest and only terrestrial class of the subphylum Chelicerata (their aquatic relatives are horseshoe crabs and sea spiders). Many arachnids are predatory, but some are parasitic or scavengers. The predators, such as scorpions and spiders, inject digestive fluids into the bodies of their prey (insects, other arthropods and small vertebrates) and then ingest the digested tissues. Mites suck blood or plant juices, or feed on dead plants and animals, depending upon the species.

The arachnid body is divided into two main parts. The unsegmented cephalothorax (comprised of a fused head and thorax), and the abdomen, which is sometimes segmented. The cephalothorax of many arachnids is covered with a hard shell called a carapace. Arachnids have six pairs of appendages. One pair of chelicerae, (feeding), one pair of pedipalps (for sensory or food manipulation), and four pairs of legs. All appendages radiate from the cephalothorax.

As a group, arachnids are misunderstood and maligned, which is unfortunate for they are invaluable in controlling the insect population. A course in predator/prey dynamics will reveal the astounding number of prey animals it takes to keep just one predator alive. Predatory arachnids are much more selective and, therefore, much more efficient ecologically and economically than pesticides.

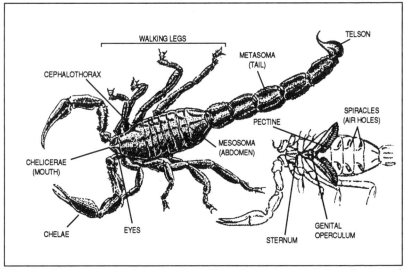

External structures of a **scorpion**. Dorsal (top) view, left and ventral (underside) view, right. *(Illustration by Jeffrey L. Martin)*

Scorpions
(Order Scorpiones)

Members of this order have a short compact cephalothorax, and an extended, clearly segmented abdomen. The last six abdominal segments are modified into a tail-like structure called the metasoma. The last segment is called the telson, and bears the sting, which is used to subdue prey as well as for self-defense. A scorpion stings by thrusting its tail forward over its head and impaling its prey, which is held in its pincers.

A scorpion's feeding appendages (chelicerae) are small and pincer-like. The pedipalps are modified into large pincers (chelae) which are used to capture prey. Just behind the fourth pair of legs are a pair of feathery sensory organs called pectines. They are unique to scorpions. It is believed they can sense low frequency ground vibrations (for locating prey) and may be chemoreceptive (for reproductive purposes). The pectines are movable, and the scorpion appears to be "feeling" the ground with them as it walks.

Scorpions are found in the South and West, occupying habitats ranging from the mountains to the low desert. They are nocturnal hunters and are active at temperatures greater than 77° F (25° C). They shun the light, hiding or burrowing during daylight hours.

Scorpions find shelter among man-made as well as natural objects. Caution should be exercised when camping or engaging in other outdoor activities, to ensure that scorpions have not found their way into footwear, clothing, sleeping bags or other equipment. If a scorpion is found in your house, check carefully for others. The easiest way to find scorpions is to shine an ultraviolet light at night. Scorpions glow brightly when UV light strikes them.

Remember, all scorpions are venomous. The sting of the bark scorpion *(Centruroides exilicauda),* in particular, can be fatal, especially to the very young and very old. Fatalities, however, are rare. If you suspect you have been stung by a scorpion, call a physician or hospital immediately. See *Symptoms & First Aid for Bites & Stings* on Page 99 for further information, or call your local Poison Control Center.

Scorpions are resilient animals whose behavior includes migrating between the outdoors and indoors regularly. Because of this they are highly resistant to improperly applied pesticides. If you suspect your house is infested, a professional exterminating service is the best solution.

Evolution has provided scorpions with venom mainly for the capture of prey. Self-defense is a secondary consideration. Scorpions would much rather be left alone. Remember that they are invaluable in controlling the insect population. They should not be molested or killed arbitrarily.

There are approximately 1200 scorpion species, 30 or so of which occur in the Southwest. While we cannot possibly cover all southwestern species in a book of this size, we have included three important species. The precautions and medical treatment for all native scorpions are basically the same.

Comparison between the potentially deadly **bark scorpion,** *(C. exilicauda)* left, the mildly venomous **devil scorpion,** *(Vaejovis spinigerus)* middle, and the mildly venomous but large **giant desert hairy scorpion,** *(Hadrurus arizonensis)* right. Note the bark scorpion's thinner tail and long slender pincers. *(Photo courtesy of Arizona-Sonora Desert Museum)*

COMMON NAME: Bark Scorpion

SCIENTIFIC NAME: *Centruroides exilicauda.*

IDENTIFICATION: Male and female are 50mm (2") to 70mm, (2 3/4") in length, including the metasoma (tail). Their color varies from brownish-yellow to tan. The chelae (pincers) are long and slender. The bark scorpion's tail is longer and thinner than that of all other Southwestern scorpions. At the base of the stinger is a small tubercle, or "tooth." The male abdomen is more slender, with a longer thinner tail, than that of the female.

RANGE: Throughout Arizona and parts of New Mexico and Mexico. A closely related species, the brown centruroides *(C.gracilis)*, occurs in Texas and east to Florida. It is similar in size and shape to the bark scorpion, but its body and pincers are dark brown.

HABITAT: Bark scorpions generally hide under tree bark, leaves and debris. They are common in mesquite, cottonwood and sycamore groves in riparian (river) areas. Bark scorpions are sometimes found in houses, especially on newly developed land. *Centruroides* scorpions are the only climbing scorpions and so are the most likely species to be found in multistory dwellings.

REPRODUCTION: During courtship, the male holds the female's pincers clasped in his own, while "dancing" back and forth. This lasts several hours. The male then deposits a spermatophore, or "sperm package," on the ground and maneuvers the female so that her genital opening will be over it so she can pick it up. The female stores the sperm until she is ready to fertilize her eggs. Females bear 18 or more live young which are miniature versions of their parents. After birth, they cling to their mother's back until they molt (around two weeks).

FOOD: Bark scorpions feed on soft-bodied insects and other arthropods.

DEFENSE: The bark scorpion's sting can be serious and is sometimes fatal. There is very little redness or swelling, but the sting is very painful. The venom is neurotoxic and affects the entire body (particularly the nervous system), causing fever, increased heart rate, restlessness and other symptoms. If you suspect you have been stung, contact your physician and the local poison control center immediately.

Close-up of the **striped centruroides'** stinger. Note the small tubercle at the base of the stinger, a trait typical of the *Centruroides* species. *(Photo by J. W. Stewart)*

COMMON NAME: Devil Scorpion, Striped-tail Scorpion

SCIENTIFIC NAME: *Vaejovis spinigerus.*

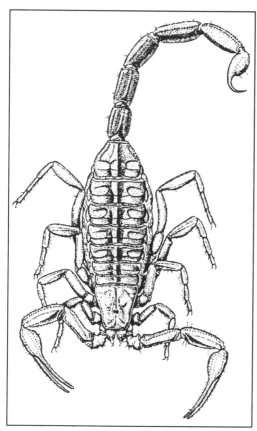

IDENTIFICATION: This stout, medium-sized scorpion, ranges in length from 50mm (2") to 70mm (2 3/4"), including the tail. Color ranges from tan to dark brown. The tail is shorter and more stocky than that of the bark scorpion and has a brownish-tan stripe on each of its sides. The pincers are shorter and more stout than those of the bark scorpion. Both sexes are similar in appearance.

RANGE: Southwestern U.S.

This illustration depicts a near term pregnant bark scorpion. *(Illustration by Jeffrey L. Martin, courtesy of Arizona-Sonora Desert Museum.)*

HABITAT: Washes, open desert, rocky terrain and along river beds. During the day, devil scorpions can be found underneath rocks, scraps of wood and dead cactus.

REPRODUCTION: The male holds the female's pincers clasped in his own, and leads her in a courtship "dance." He then deposits a stalk-like spermatophore, or "sperm package," on the ground and maneuvers the female so that her genital

The **devil scorpion** is a mildly venomous species found in the Southwest. It is about the same size as the bark scorpion, but has a thicker tail, abdomen and pincers. *(Photo courtesy of the Arizona-Sonora Desert Museum)*

opening will be over it so that she can pick it up. The female bears live young and carries them on her back for 10 to 15 days until their first molt. The young scorpions scatter, moving on to fend for themselves.

FOOD: Small, soft-bodied insects.

DEFENSE: Mildly venomous. Sting causes swelling and redness at the site. Wash the wound with soap and water, and apply an antiseptic. Consult your physician.

COMMON NAME: Giant Desert Hairy Scorpion

SCIENTIFIC NAME: *Hadrurus arizonensis*

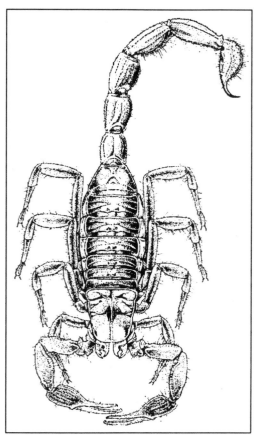

IDENTIFICATION: The giant desert hairy scorpion is the largest of scorpion species found in the Southwest. Adults can reach 14cm (5 1/2") in length. Their legs, pedipalps and tail are yellowish brown or greenish, and bear erect brown hairs. the cephalothorax and abdoman are dark grey or brown, rimmed with yellow. Sexes are similar in appearance.

RANGE: Throughout the Southwest.

HABITAT: Desert washes and open areas of desert floors.

The **giant desert hairy scorpion** is the largest scorpion species in the Southwest. *(Illustration by Jeffrey L. Martin, courtesy of Arizona-Sonora Desert Museum.)*

REPRODUCTION: The male holds the female's pincers clasped in his own, and leads her in a courtship "dance." He meanwhile deposits a stalk-like spermatophore, (sperm package) on the ground. The male then maneuvers the female so that her genital opening is positioned above the spermatophore. The female bears live young and carries them on her back for about 20 days. After their first molt, young scorpions scatter and live on their own.

FOOD: Soft-bodied insects.

DEFENSE: The giant desert hairy scorpion looks dangerous because of its large size, but it is only mildly venomous. Its sting causes swelling and redness at the site, but usually no systemic reactions. Wash the wound with soap and water, and apply an antiseptic. Consult your physician.

Scorpion-like Arachnids

The four following arachnid orders contain species which resemble scorpions or spiders. Despite the menacing appearance of some of these creatures, they are not dangerous. In fact, they are beneficial to humans because they all eat insects. Some of the selected arachnids which follow are not venomous at all and are only included in this book because they are often mistakenly believed to be venomous or dangerous.

Pseudoscorpions
(Order Pseudoscorpionida)

Pseudoscorpions are tiny arachnids that resemble true scorpions. They possess pincers, but lack the stinging tail. Some 2000 species occur worldwide. They are extremely small, measuring 8mm (5/16") or less. Pseudoscorpions are common but rarely seen because of their small size and their preference for living among debris. The pincers, which have venom glands, are too small to impale anything but the tiny invertebrates upon which they feed.

COMMON NAME: Pseudoscorpions

SCIENTIFIC NAME: A number of genera occur in the Southwest, including *Dinocheirus* and *Chelifer*.

IDENTIFICATION: Small. Pseudoscorpions range from 2mm (1/16") to 8mm (5/16") in length. Their color ranges from yellowish-brown to blackish-blue. Males and females are similar in appearance. These arachnids resemble true scorpions, possessing enlarged pincers, but no tail or stinger. Pseudoscorpions have silk glands in their chelicerae (jaws) with which they spin cocoons for molting and nesting.

RANGE: Southwestern deserts to forest mountain elevations. Common but reclusive. *Chelifer cancroides* can sometimes be found in houses. However, they are commonly in compost heaps.

HABITAT: Pseudoscorpions live in leaf mold, decaying vegetation, under the soil, underneath stones and tree bark, in rotting cactus and in the nests of small mammals. When populations become too dense, pseudoscorpions "hitch a ride" by clinging to the legs of flying insects such as bees, beetles and flies and settle in new areas.

REPRODUCTION: The male deposits a stalk-like spermatophore on the ground. The female is attracted by the scent and then is maneuvered over the spermatophore by the male so that she can pick it up. She then builds a silk-lined nest and deposits up to 50 eggs in a brood pouch, which remains attached to her ventral side (underbody). The eggs develop in the pouch. Neonates undergo two molts, one molt before hatching, and another before emerging from the pouch. In some species, the embryos are fed a "milk" secreted by the ovaries of the mother.

FOOD: Small flies, bark lice, ants, mites, small earthworms.

DEFENSE: Pseudoscorpions have venom glands located in their pincers. Venom is used to subdue prey only. The pincers are far too small to "bite" humans.

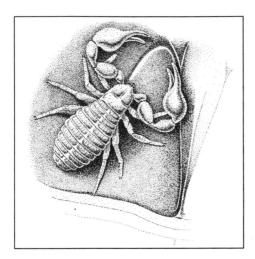

Some species of **pseudoscorpions** live under the elytra (thick or leathery front wing) of beetles, feeding on mold and parasitic mites. *(Illustration by Jeffrey L. Martin)*

Windscorpions
(Order Solpugida)

There are about 900 solpugid (or solifugid) species, 100 or so of which occur in the Southwest. Windscorpions are so-called because they "run like the wind." Some species are diurnal (active during the day) and are known as sunspiders. Sopulgids are small to large arachnids with spider-like legs and a clearly segmented abdomen like that of a scorpion. They have neither pincers, nor a stinging tail. The windscorpion's prominent possession is a pair of proportionately large forward-facing pincer-like chelicerae. They are non-venomous, but can impart a painful bite. Windscorpions are solitary and many species hunt at night for insects and small vertebrates.

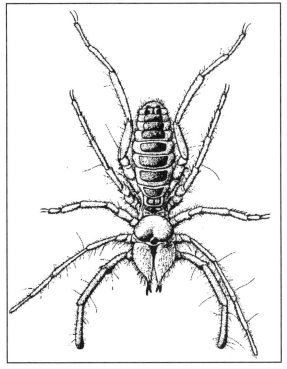

Windscorpion. Note the large chelicerae, centered eyes, and pedipalps which are used as feelers. *(Illustration by Jeffrey L. Martin, courtesy Arizona-Sonora Desert Museum)*

COMMON NAME: Windscorpion, Sunspider

SCIENTIFIC NAME: *Eremobates* is one of a number of genera that occur in the southwest.

IDENTIFICATION: 10mm (3/8") to 50mm (2"). Yellowish-brown in color, with large forward-facing chelicerae (mouthparts). Its pedipalps are as long and as stout as its legs. They are used as feelers, as are the first (and much thinner) pair of "walking legs." Sexes are similar in appearance.

RANGE: Around 100 species are found throughout the Southwestern United States.

HABITAT: Nocturnal. Open deserts and semiarid lands. Hides by day in debris, under bark of dead trees or in underground burrows.

REPRODUCTION: A sperm droplet is transferred by the male into female's genital opening (called the "gonopore") by means of the male's chelicerae. The female then digs a burrow in the soil, lays 50 to 100 eggs and stands guard over them until they hatch. She will stay with the neonates for several weeks, providing food for them.

FOOD: Small soft-bodied insects, and small vertebrates, such as lizards.

DEFENSE: Windscorpions lack venom glands. However, their bites can be very painful. Wash the wound with soap and water, and apply an antiseptic.

Whipscorpions
(Order Uropygi)

These arachnids are named for the whiplike tail located at the tip of the abdomen. About 130 species of whipscorpions can be found in tropical and semitropical regions worldwide. As an order these arachnids range in size from minute to very large. The largest member of this group, the whipscorpion, resides in the South and Southwest. Measuring up to 80mm (3") in length (not counting its whip), it resembles a scorpion. Whipscorpions possesses a pair of pincers which are non-venomous, but can

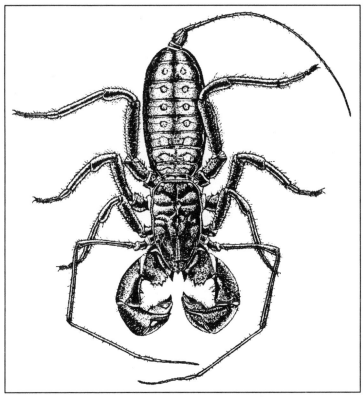

Male **whipscorpion.** Compare with female on next page. *(Illustration by Jeffrey Martin, courtesy Arizona-Sonora Desert Museum)*

deliver a painful pinch to humans. The tail has no sting, but they can eject a stream of acidic liquid from it in self-defense.

COMMON NAME: Whipscorpion, Vinegaroon

SCIENTIFIC NAME: *Mastigoproctus giganteus*

IDENTIFICATION: These arachnids resemble scorpions but differ from them in the shape of their pedipalps. They range in size up to 80mm (3") long, not including the slender whip-like "tail." Brown to black in color, their first pair of walking legs are thin and used as feelers. Sexes are similar in appearance.

RANGE: From coast to coast in the South and Southwest.

HABITAT: Whipscorpions are nocturnal. They hide by day

Despite its appearance, the **whipscorpion** is not dangerous. Because it is an insectivore, it is actually beneficial to man. *(Photo courtesy of Arizona-Sonora Desert Museum)*

under stones, rotting wood and in burrows, and indoors in dark corners.

REPRODUCTION: Whipscorpions perform a courtship dance similar to that of scorpions. The male releases a spermatophore which he inserts into the female's genital opening (gonopore) with his pincers. The female carries 25 to 35 eggs in a membranous sac below her abdomen. She remains inside a small shelter until they hatch. Neonates ride on their mother's back. After molting several times, the young disperse and the female dies.

FOOD: Whipscorpions feed on slugs, worms and arthropods.

DEFENSE: When alarmed, the whipscorpion elevates and directs its whip, which sprays a mixture of acetic and caprylic acids. This ejection may reach distances of 80cm (31.5") and can cause eye irritation or a burning sensation if it gets into an open wound. The caprylic acid is capable of dissolving the exoskeleton (shell) of insects and other arthropods. Because this mix smells like vinegar, whipscorpions are sometimes called vinegaroons.

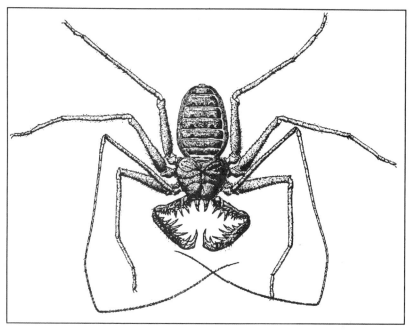

Tailless whipscorpions look dangerous, but as they lack venom glands, are really quite harmless. *(Illustration by Jeffrey L. Martin, courtesy Arizona-Sonora Desert Museum)*

Tailless Whipscorpions
(Order Amblypygi)

This order is comprised of about 60 species, which live in mostly humid areas in Africa, Asia, and the Western Hemisphere. Although their range is limited in the Southwest, tailless whipscorpions can be found in homes, particularly those in newly developed areas. They are generally spider-like in appearance, but they differ structurally from spiders. The abdomen is clearly segmented. The first pair of walking legs are long, whip-like (hence, the common name) and are used as feelers. The pedipalps have been modified into heavy, spiny claws that are similar in form and function to the grasping legs of the praying mantis. They are used to capture and hold the insects upon which they feed. Despite their formidable appearance, these arachnids are not venomous.

COMMON NAME: Tailless Whipscorpion

SCIENTIFIC NAME: *Paraphrynus mexicanus*

IDENTIFICATION: Body length to 2 cm (3/4"), flattened and spiderlike in appearance, but the pedipalps are enlarged and spiny. The first pair of walking legs are slender and whiplike and are used as sensory organs. All of its legs are quite long which makes it look larger than it is. Tailless whipscorpions are pale brown to blackish in overall color and sometimes have spots. This arachnid retreats with a crablike sideways motion when disturbed.

RANGE: Mexico north to central Arizona. Other species occur in southeastern United States.

HABITAT: Most Amblypygids prefer warm humid regions, but P. *mexicanus* can also be found in the arid Sonoran Desert. We have found specimens in San Carlos, Sonora, Mexico (a desert area which is humid because of its proximity to the Gulf of California) and in Phoenix, Arizona (which is relatively drier). This nocturnal creature hides under rocks and in crevices in the bark of trees by day. It is frequently found indoors in dark places (shower walls, under sinks, in cellars etc.).

REPRODUCTION: Both sexes touch each other with sensory legs during courtship. The male deposits a spermatophore stalk and guides the female over it by using his pedipalps. The female lays up to 60 eggs and carries them in a hardened sac under her abdomen for 4 to 6 days. The hatchlings then emerge and ride on her back until their first molt (usually within a week) and then disperse.

FOOD: Tailless whipscorpions hunt insects and other arthropods during the night hours. Prey is located with sensory legs then pounced on and grasped with the pedipalps.

BITE: Tailless whipscorpions are non-venomous and do not bite or sting.

Spiders
(Order Araneae)

Though superficially dissimilar to scorpions, spiders possess the basic anatomical structures common to all arachnids. These features are merely modified in shape, proportion and purpose. There is also much variety between species, as well as between sexes within each species. Southwestern spiders range in size from 3mm (0.1") or smaller, up to 10 cm (4") in length. Their abdomens may be rectangular, ovoid, globular or triangular. Spiders range in color from drab grey and brown, to bright yellow, red, orange, green, silver, or striking and beautiful combinations of color. Females are generally larger than males — many times larger in some species.

In members of this order, the cephalothorax (head and thorax combined) is usually equal to or smaller than the abdomen. These are separated by a narrow waist called a pedicel. The cephalothorax of nearly all spiders bears eight eyes, which are arranged differently in each family. Some spiders, such as brown spiders *(Loxosceles)* have six eyes. Others have two or none. The cephalothorax also bears all feeding, sensory and locomotive appendages.

The chelicerae are fang-like and are used to capture and inject venom into prey, such as insects, and small vertebrates. The pedipalps are modified into leg-like feelers, which are used

Scanning electron micrograph of a **black widow spiderling** after its first molt. The chelicerae, pedipalps and eyes are plainly visible. Note sensory hairs on the body and all appendages. *(Photo by Norman Grim, Ph.D.)*

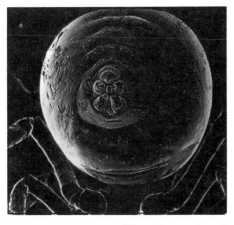

Scanning electron micrograph. Spinnerets of a **black widow spiderling** (after the first molt). *(Photo by Norman Grim, Ph.D.)*

to sense prey as well as obstacles. Spiders have four pairs of legs that are used for running or jumping.

The abdomen, which is unsegmented, bears six to eight spinnerets at the rear end. These organs produce silk strands which serve a variety of purposes (capture of prey, door seals, etc.). Many spiders play out safety lines, called "draglines," as they walk. Should a spider lose its footing, or need to make an escape, the dragline can become a lifesaver.

Spiders employ various methods in capturing prey, depending on the species. Orb-web spiders build sticky webs of complex geometric beauty, designed to capture winged insects in flight. Wolf spiders and jumping spiders pounce on their prey. Crab spiders (among the most brightly colored spiders) commonly sit in flowers and wait in ambush for insects which come to feed on nectar and pollen. Spitting spiders release venom and glue from glands in their chelicerae. Other methods include various silken snares.

Because spiders are predatory, mating can be a dangerous process, particularly for the male. It is the male which must approach the female. He is smaller than she (many times so in a great number of species) and a potential meal as well as mate.

Complex courtship rites have developed to determine the sexual maturity and receptivity of the female. These rites vary according to the type of spider. Jumping spiders have excellent vision and so take visual cues (size, color, leg movement). Tugging at webs to send a mating signal is the tactic of orb web spiders.

Wolf spiders test the draglines of females for pheromones (natural "perfumes"), and can determine the species, as well as sexual maturity by the scent.

The pedipalps of male spiders have special reservoir organs with which sperm is transferred to the female. Before seeking a mate, the male spider uses these to collect sperm from a genital pore located at the anterior end of its abdomen. During mating the male places the sperm in the female's reproductive opening. She stores the sperm until she is ready to fertilize her eggs.

Nearly all spiders are venomous to some degree. Two species are particularly infamous for their toxicity: the black widow *(Latrodectus hesperus)* and the brown or violin spider *(Loxosceles arizonica)*.

Evolution has provided spiders with venom primarily for the capture of prey. Self-defense is a secondary consideration. Spiders would much rather be left alone and almost invariably retreat at the first sign of danger. Remember that they are invaluable in keeping a check on the insect population. Should you find a spider indoors (and if this is unacceptable), consider freeing it outdoors, if you can do so without risking a bite. Spiders are highly beneficial and should be appreciated as such.

Note: All dimensions of species in this chapter refer to the combined length of the cephalothorax and abdomen. Length of legs (which are considerably longer) is not included.

The appendages at the rear of the **tarantula's** abdomen are spinnerets. *(Photo courtesy Teaching files, Dept of Entomology, Texas A&M University)*

The male **tarantula** wanders the desert floor in late summer in search of a mate. *(Photo by Terry Christopher)*

COMMON NAME: Tarantula

SCIENTIFIC NAME: Family Theraphosidae

IDENTIFICATION: Large, hairy spiders. Male is 5 to 6 cm (2" to 2 1/2"). The female can reach 10 cm (4"). Both have a dark brown or grey cephalothorax. The male has thinner, longer legs. The first pair bears hooks used to hold the female's fangs during mating. Tarantulas have 8 eyes which are grouped in a tight cluster front and center at the top of their cephalothorax.

RANGE: Southwestern deserts.

HABITAT: Tarantulas are found in washes and open deserts.

REPRODUCTION: The mating season lasts from June through October. This is when tarantulas are the most visible, crossing roads and open desert. At an age of 8 to 10 years, the male goes through a final molt, and wanders from its burrow in search of a mate. The female has an established burrow prior to mating, in which she spins a large silk sheet onto which she will lay her eggs. Female tarantulas may live up to 20 years, but the male dies within a year of its last molt.

Tarantula in threatening defense position. The top surface of the abdomen bears brittle hairs. The spider can disperse these into the air by rubbing them rapidly with its hind legs. *(Photo courtesy Teaching files, Dept of Entomology, Texas A&M University)*

FOOD: Arthropods, and small vertebrates (mice, lizards, snakes).

DEFENSE: Toward man, the tarantula is usually a very timid, secretive and harmless species, but it will bite if provoked. Because of its large fang-like chelicerae, the bite can be very painful but it is not life threatening. Another method of defense is the cloud of fine hairs that it can create by rubbing its hind legs over the bristles on the top of its abdomen. These hairs can be extremely irritating to the eyes and mucous membranes of a would-be predator.

COMMON NAME: Wolf Spider

SCIENTIFIC NAME: Family Lycosidae

IDENTIFICATION: From 7mm to 35mm (3/8" to 1 3/8") in body length. Males are moderately smaller than females. Cephalothorax and abdomen are grey, brown or a combination of both. Many are striped or spotted. Most wolf spiders have

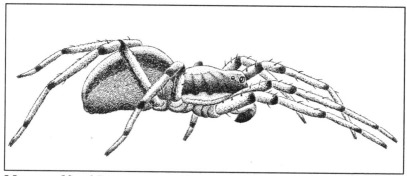

Many **wolf spiders** are striped or spotted, with dense short hair on their bodies and appendages. Note the large pedipalps on this male wolf spider. *(Illustration by Jeffrey L. Martin)*

dense short hair, as well as some longer hairs on their bodies and appendages. The pedipalps of the male are larger and hairier than those of the female. Six forward facing eyes (two large, four small) and an eye on either side provide excellent eyesight.

RANGE: Throughout the United States.

HABITAT: In the Southwest, washes, open desert, flatlands, open fields, forests. Wolf spiders can sometimes be found sunning themselves by day.

REPRODUCTION: In courtship, the male rhythmatically waves his pedipalps in front of the female to attract her attention. Using his pedipalps, the male transfers sperm to the female's genital opening, located on the underside of her abdomen. After mating, the female produces an egg sac which she carries at the rear of her abdomen until the eggs hatch. She then carries the spiderlings on her back until their first molt.

FOOD: Wolf spiders actively hunt at night for small, soft-bodied insects. With a few exceptions, they do not use a web to capture their prey.

DEFENSE: The bite of North American wolf spiders is only mildly venomous. Wash the wound with soap and water, and apply an antiseptic. Consult a physician in the event of unusual reactions, such as infection, or allergic reaction.

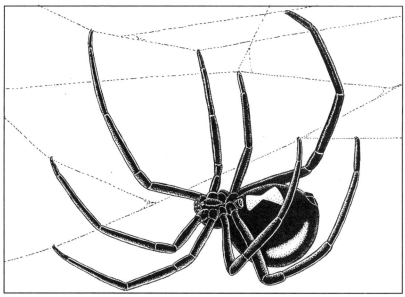

Distinctive identifying feature of female **black widow** is the red hourglass mark on its underside. Some also have brown or white stripes and spots on the abdomen. *(Illustration by Jeffrey L. Martin)*

COMMON NAME: Black Widow Spider

SCIENTIFIC NAME: *Latrodectus hesperus*

IDENTIFICATION: Female is glossy black. Abdomen globular, with a red hourglass on the underside. Her body can reach 12mm (1/2") in length. Males are much smaller and are grey in color with stripes or spots. Spiderlings are orange, brown and white and change coloring with each molting.

RANGE: Throughout Arizona, New Mexico, Nevada, Utah and Southern California. Very similar to eastern species — *L. mactans.*

HABITAT: Black widows are found in rock crevices, wood piles, overhangs, dark garages, basements, stables and abandoned rodent holes. Spins an irregular web of silk strands which are heavy and strong.

REPRODUCTION: The male transfers sperm to the female's genital opening. The female stores the sperm and can produce several egg sacs from only one mating. The female spins an egg sac in which she lays 150 eggs or more. She places it in the web and guards it tenaciously until the eggs hatch in about 30 days. The spiderlings climb plants or other high structures on a windy day and spin a long strand of silk. When the breeze catches the silk they "balloon" along with it to a new home. The female occasionally eats the male after mating.

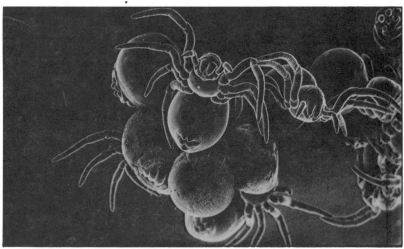

Scanning electron micrograph. Newly hatched **black widow spiderlings.** *(Photo by J. Norman Grim, Ph.D.)*

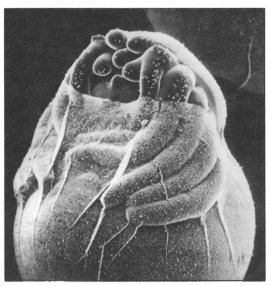

Scanning electron micrograph. **Black widow spiderling** in its egg. *(Photo by J. Norman Grim, Ph.D.)*

FOOD: Insects

DEFENSE: Highly venomous — a bite *can* be, but very rarely is fatal. The black widow's neurotoxic venom affects the nervous system. In humans, a black widow's bite can cause systemic pain, nausea, muscle spasms, and respiratory paralysis. There is no first aid procedure for bites from this spider. Consult a physician or your poison control center immediately.

Female **black widow** with egg sac. *(Photo by J. W. Stewart)*

COMMON NAME: Brown Spider or Violin Spider

SCIENTIFIC NAME: *Loxosceles arizonica*

IDENTIFICATION: Brown spiders can grow to 9mm, (5/16") in length with a leg span of 25mm (1"). They are tan to brown in color with violin-shaped marking on the cephalothorax (the neck of the violin points to the spider's abdomen). They have three pairs of eyes situated at the base of the "violin."

RANGE: Southwestern United States. A similar southwestern species, the desert loxosceles *(Loxosceles deserta)* is lighter in color and much less venomous. The closely related and

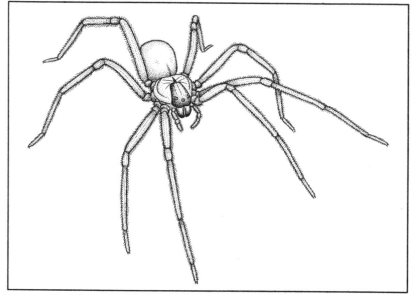

Brown spider, or **violin spider.** Note violin shaped marking on cephalothorax and three pairs of eyes. Most spiders have eight eyes, but the brown spider has only six. *(Illustration by Jeffrey L. Martin)*

highly venomous, brown recluse spider *(Loxosceles reclusa)* is found throughout the southeastern United States.

HABITAT: Brown spiders live under rocks, pieces of wood, and in nests and burrows of other animals. As they prefer dark places—basements and garages can become their home.

REPRODUCTION: Female brown spiders build egg-sac cocoons during the warmer months.

FOOD: Constructs a sticky, sheet-like web for capturing small insects.

DEFENSE: Highly venomous, but nonaggressive. A bite from this spider results in the death of tissue surrounding the initial wound. The severity and spread of tissue degeneration varies from incident to incident, and may depend on the amount of venom injected, and the victim's immune response. Bites most often occur because the spider has hidden itself in bedding or clothing, and is subsequently disturbed. There is no first aid procedure for a bite from this spider. Consult a physician as soon as possible.

The **crab spider** is one of the many beautiful species of spiders found in the Southwest. *(Photo courtesy Entomological Society of America)*

COMMON NAME: Crab Spider

SCIENTIFIC NAME: Family Thomisidae

IDENTIFICATION: Typically flattened and crablike in appearance and movement. First and second leg pairs are much longer and more stout than the third and fourth and are held out in a crablike fashion. In females, the abdomen is flattened, ovular and larger than the cephalothorax. The abdomen of the male is generally smaller. Many species have ornaments, such as horns or bumps, on the abdomen and cephalothorax. This spider walks sideways and backwards as well as forward. Crab spiders range in size from 3 mm (1/10") to the large tropical species, which measure as much as 20mm (7/8").

RANGE: Worldwide.

HABITAT: Mountain and lowland forests, open fields, semi-arid desert.

REPRODUCTION: In some species, the male may wrap the female in silk before mating. After mating, the female spins a silk cocoon in which she deposits her eggs. She defends the cocoon vigorously and does not feed. She usually dies before the eggs hatch.

FOOD: Crab spiders do not use webs, but employ a camouflage and ambush hunting strategy. They hide on tree bark, on leaves and flower petals, etc. waiting for insects (including those larger than the spider itself!). Some crab spiders can change their colors to match the plant they inhabit. This feat takes several days. On one occasion in the Coconino National Forest, Arizona, we observed a beautiful bright yellow specimen (possibly *Misumena vatia*) on a flower of precisely the same color. The spider would have gone undetected if not for the black fly it had just caught and was in the process of consuming.

BITE: The bite of crab spiders is only mildly venomous. Wash the wound with soap and water, and apply an antiseptic. Consult a physician in the event of unusual reactions, such as infection or allergic reaction.

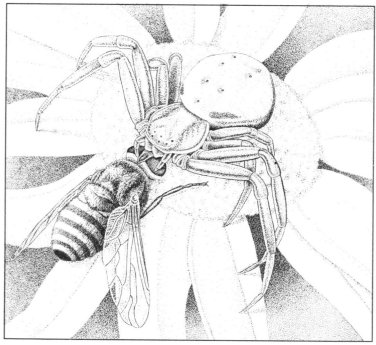

The **goldenrod crab spider** *(Misumena vatia)* can assume the color of a flower in a few days. This specimen is feeding on a freshly caught bee fly. *(Illustration by Jeffrey L. Martin)*

Ticks & Mites
(Order Acari)

This arthropod order is extremely vast and successful. There are over 30,000 known species, with perhaps a million as yet undescribed by science. Mites can be found nearly everywhere on earth, and can live under the most extreme conditions. They are diverse in habit and habitat. They live on land, in freshwater, and even in the ocean. Due to their small size (some are nearly microscopic), mites are capable of exploiting niches unavailable to larger animals. For example, there are mites which form symbiotic (coexistent) relationships with ants, centipedes, beetles, and other arthropods. While some of these mites are parasitic, others are merely "hitching a ride," which aids in the dispersal of the species.

The acarid cephalothorax and abdomen are completely fused. Abdominal segments are absent or at least not apparent. The mouthparts (capitulum) are used for a variety of feeding strategies. While most mites ingest liquids, some species can feed on solid food particles, an uncommon characteristic among arachnids. Thousands of mite species are scavengers, feeding on decomposing plants and animals. Many are beneficial predators, and have been used to control agricultural pests, such as other mites.

Ticks are the largest of the mites. Individuals of the largest species, when engorged with blood easily exceed an inch (2.54 cm) in length. Ticks infest wild animals, livestock, and man.

Ticks, in general, are vectors of a variety of bacterial and viral infections. The deer tick *(Ixodes pacificus)* is a small Western species. It has been described as a vector of the microorganism, *Borrelia burgdorferi,* which causes Lyme disease. The Rocky Mountain Wood Tick *(Dermacentor andersoni)* is responsible for Rocky Mountain spotted fever.

Humans become infested with ticks as secondary hosts. Hatchling ticks find a host and feed after which they drop off and molt. They then climb a plant or wall, etc. and await a new host to brush by. This is the stage at which a human is most likely to pick up a tick.

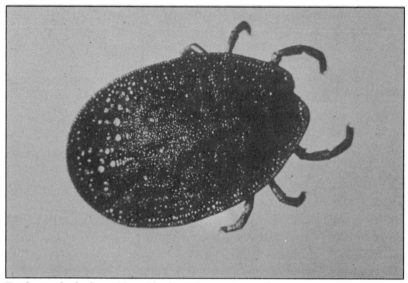

Pock marks help to identify the **adobe tick**. *(Photo courtesy Teaching Files, Dept of Entomology, Texas A&M University)*

COMMON NAME: Adobe Tick, Fowl Tick

SCIENTIFIC NAME: *Argas sanchezi*

IDENTIFICATION: Adult females average 8.5mm (5/16"). Males average slightly smaller in size, to 6.5mm (1/4"). Oval and flat in shape with wrinkled and sac-like abdomens. Both are pocked with small hair-bearing "buttons." They are light-red to dark brown in color.

RANGE: Southwestern deserts.

HABITAT: Adobe ticks are found in farm buildings, ranches and in chicken coops.

REPRODUCTION: Egg laying is always preceded by a blood meal. Egg masses of up to 100 eggs hatch in 12 to 28 days.

FOOD: Blood of birds and mammals, including human.

BITE: The saliva of this tick causes inflammation at the site of the bite. Small granular tumors may form, but disappear within two to three weeks. Wash the wound with soap and water and apply an antiseptic. Consult a physician in the event of unusual reactions.

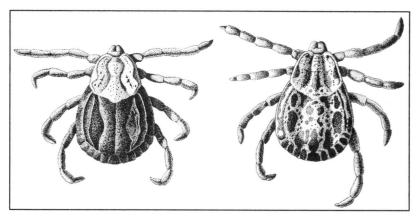

Rocky Mountain wood tick.   Female, *left,* and male, *right.*
(Illustration by Jeffrey L. Martin)

COMMON NAME: Rocky Mountain Wood Tick

SCIENTIFIC NAME: *Dermacentor andersoni*

IDENTIFICATION: Up to 9mm (3/8") long (engorged females become larger). Oval-shaped. The female is reddish-brown, with a white cephalothorax. Males are white with dark brown spots. Engorged females are greyish.

RANGE: Throughout North America.

HABITAT: Wood ticks are found in brushy areas and yards where animals are present.

REPRODUCTION: The female lays up to 6,000 eggs in a month's time, then dies. Within two months the eggs hatch. This is a three host tick. One each in the larval, nymphal and adult stages. After feeding, it drops off to molt and find a new host.

FOOD: The blood of mammals, including humans.

BITE: Can cause tick paralysis and may carry Rocky Mountain spotted fever, which can be fatal to humans. Consult a physician if you suspect that you have been bitten by any tick species.

Brown dog tick. Larger tick is a female engorged with blood; male is on the right. *(Photo courtesy of Entomological Society of America)*

COMMON NAME: Brown Dog Tick

SCIENTIFIC NAME: *Rhipicephalus sanguineus*

IDENTIFICATION: Adult males can reach 4mm (3/16") in length. Both sexes are reddish-brown and have small pits on their backs. Engorged females turn a bluish-gray in color and can become 12mm (1/2") in length.

RANGE: Throughout the Southwest.

HABITAT: Can be introduced to urban dwellings by pets.

REPRODUCTION: Female drops from the host, lays up to 3,000 eggs and then dies. The eggs hatch within 2 months and the larvae climb walls, trees, etc. to wait for a host. They then feed, molt into nymph stage, feed again and then drop off to become adults.

FOOD: Dog parasite.

BITE: Can cause tick paralysis in dogs and, (very rarely) in humans.

NOTE: Because of its tendency to deposit egg masses in well hidden places (e.g. behind furniture), the brown dog tick can easily infest homes. Diligent dog hygiene and house vacuuming (walls, floors, curtains, etc.) are necessary. Consult a pest control expert or veterinarian.

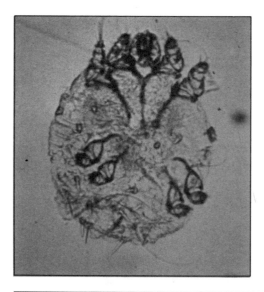

The **scabies mite** viewed through a microscope. *(Photo courtesy Teaching Files, Dept of Entomology, Texas A&M University)*

COMMON NAME: Scabies Mite

SCIENTIFIC NAME: *Sarcoptes scabiei*

IDENTIFICATION: Scabies mites are round, translucent, off-white in color and have very short legs. They average 0.20mm (8/1000") to 0.45mm (17/1000") in length.

RANGE: Throughout the world.

HABITAT: Animal parasite.

REPRODUCTION: Females lay eggs in burrows made in the host's skin. Larvae and nymphs move over skin surface, sometimes "nesting" in hair follicles.

FOOD: Blood of warm blooded animals, including man.

BITE: Scabies mites burrow beneath the skin within a few minutes of contact. Symptoms include itching and irritation of the skin. Scratching can cause the spread of infection to hands. Once diagnosed, a scabies infestation can be easily treated. Topical formula is available by prescription. Treat all family members and sterilize clothing and bedsheets. Mites cause mange in livestock and domestic animals.

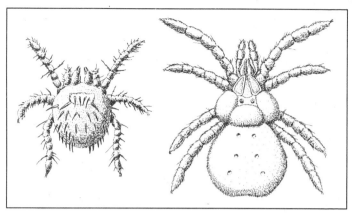

Chigger mite (*Trombicula* sp.),larva, *left,* and adult. *(Illustration by Jeffrey L. Martin)*

COMMON NAME: Chigger Mite, Red Bug

SCIENTIFIC NAME: *Trombicula* sp.

IDENTIFICATION: Red, hairy mite. Adult can grow to 3mm (1/10") in length.

RANGE: Throughout the United States.

HABITAT: Brushy areas or grasslands.

REPRODUCTION: The eggs are laid on moist soil and hatch in four to five days. Larval mites climb tall grasses, waiting for a host to continue their reproductive circle.

FOOD: Adults feed on insects and insect eggs. Larvae are parasitic, feeding on the skin of amphibians, reptiles, birds and mammals, including humans. Larvae drop from tall plants onto an animal host for a three to four day feeding session then drop off to molt.

BITE: Intense itching is experienced within three to six hours of the initial bite. A red spot (the chigger larva) can be seen at the center of the inflamed area. A heat sterilized needle may be used to remove them or Calamine® lotion and antiseptics can be used to relieve the itching caused by these mites. Treat affected areas frequently. Consult a physician in the event of infection.

Myriapods

Centipedes — Class Chilopoda

The bodies of centipedes are elongate and flattened with multiple segments. Each segment bears one pair of legs (between 15 to 181 pairs, depending upon the species). Centipedes have one pair of antennae. Most have simple eyes, some have complex eyes, while others have none at all. The most important senses are touch and smell. Centipedes are predators, feeding on insects and small vertebrates, depending upon the species. They subdue their prey with a mild venom issued from fangs called gnathopods, which are actually specially modified legs.

As a rule, the less legs a centipede has, the faster it can run. Three out of four centipede orders are designed for running. One order lives in the soil and gets around by expanding and contracting its body, in much the same way as an earthworm. Of the approximately 3000 known centipede species, most are rather small, falling somewhere in the 3cm to 6 cm (1 1/8" - 2 1/4") range. However, some species grow to an impressive size. *Scolopendra gigantea,* of tropical Central and South America, reaches an astonishing 26cm (10 1/4") in length.

Centipedes are global in distribution, occurring most commonly in the temperate and tropical regions. They are secretive, living under rocks, in burrows or among soil debris, and prefer humid environments. They occur at a wide range of elevations. Mountains, forests, meadows and deserts are all inhabited by these arthropods. Some coastal centipede species even make their homes in the littoral zone (area between high and low tides). Littoral centipedes survive for hours completely immersed in water.

Centipedes of the Southwest prefer to hunt at night. They can, however, be seen during daylight hours when the weather is cloudy or rainy. They are occasionally encountered indoors, particularly in newly developed areas. Since they are not dangerous and are beneficial insectivores, they should be released outdoors away from human habitation.

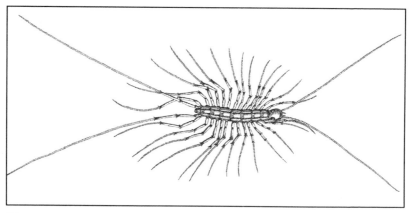

The **house centipede** *(Scutigera coleoptrata)* is the fastest runner of all the centipedes. *(Illustration by Jeffrey L. Martin)*

COMMON NAME: House Centipede

SCIENTIFIC NAME: *Scutigera coleoptrata*

IDENTIFICATION: Both males and females are brownish-yellow in color. Their average length is 35mm (1 3/8"). Antennae are longer than the body. They have fifteen pairs of long and slender legs, each pair longer than the pair before it. The fifteenth pair resembles antennae. They have a rounded head with compound (multifaceted) eyes. Because their back plates are hooked together, they can hold their back rigid and well off the ground making it possible for them to run very fast.

RANGE: United States and Europe.

HABITAT: This species avoids dehydration by hiding during the day in dark, humid places, such as under stones, bark and debris. Of all centipedes, the House Centipede is the one most likely to be found in human habitations.

REPRODUCTION: The male and female touch each other with vibrating antennae and form a circle. The male positions himself beneath the female and rocks up and down. This is repeated numerous times, until he produces a small oval spermatophore (sperm packet) on the ground. He then maneuvers the female over the spermatophore so that she can pick it up with the posterior end of her body.

The average female lays 60 eggs, but the number is occasionally double this amount. Eggs, which measure 1.2mm (3/64"), are laid one at a time. Each egg is coated with a secretion from the female's posterior. The egg is then pushed about in soil until it resembles a tiny lump of dirt. The egg is then dropped into a crevice in the soil and abandoned. Hatchlings have four pairs of legs and they gain additional pairs with subsequent molts. House centipedes molt six times before they are completely mature.

FOOD: Flies, spiders, and other small soft-bodied arthropods. Unlike other centipedes, this one hunts out in the open.

BITE: Centipedes are nonaggressive and their venom is mild. The bite of a centipede is painful at worst. The afflicted area may remain tender for a few weeks and heal slowly. Tetanus and other infections are the chief concerns with centipede bites. Allow the wound to bleed, wash with soap and warm water and apply an antiseptic. Consult a physician for information regarding tetanus shots.

The **giant desert centipede** is the largest North American centipede and is found in the Southwestern deserts. *(Photo courtesy Arizona-Sonora Desert Museum)*

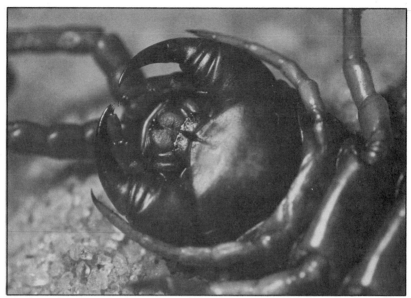

Underside of head of the **giant desert centipede.** Note the fang-like gnathopods, which are actually modified legs. They contain poison glands and are used for capturing prey. *(Photo by J. W. Stewart)*

COMMON NAME: Giant Desert Centipede

SCIENTIFIC NAME: *Scolopendra heros*

IDENTIFICATION: The largest North American centipede, averaging 17cm (6 11/16") in length. It is mostly brown in color with bluish head and tail segments. Legs stemming from blue tail segments may also be blue. This centipede has 21 pairs of legs, the first pair being slightly shorter than the width of the centipede's body. Each pair is slightly longer than the pair before it. The 20th pair is nearly twice as long as the body width. The last pair is more than double the body width, very heavy, and facing to the rear. This gives the centipede the appearance of having a head on each end. Their flattened heads have four simple (single) eyes on each side. Sexes are similar in appearance.

RANGE: Throughout Southwestern United States.

HABITAT: Arid desert. This centipede retreats to a burrow by day, especially when weather is hot and dry. On cloudy or rainy days, it will become active and leave the burrow during daylight hours.

REPRODUCTION: Male and female line up side by side or in a circular formation, head to tail. Each taps the other's posterior with its antennae, and the pair move in a circle. This courtship process can last for hours. The male then spins a small silk web on the ground and places a spermatophore (sperm packet) on the web. He signals the female and directs her to the spermatophore, which she then picks up with her posterior end.

The female lays a sticky egg-mass of approximately twenty eggs in a hollowed out cavity in dead wood or cactus. She guards her brood by coiling her body around the egg-mass. The young centipedes hatch with their full compliment of 21 pairs of legs. She guards her young until they are ready to fend for themselves.

FOOD: This mostly nocturnal hunter is large enough to prey on vertebrates such as small lizards and baby mice, but more commonly takes soft-bodied insects and other arthropods. It captures its prey with fang-like modified legs called gnathopods. Poison glands in each of the gnathopods secrete a mild venom which subdues the prey animal.

BITE: Centipedes are nonaggressive and their venom is mild. Their bite is painful at worst. This is especially true of the giant desert centipede because of its large size. After 15 minutes the pain begins to subside and is gone within a day. Some swelling occurs, but this is minor. The afflicted area may remain tender for a few weeks and heal slowly. Tetanus and other infections are the chief concerns with centipede bites. Allow the wound to bleed and wash with soap and warm water. Apply an antiseptic. Consult a physician for information regarding tetanus shots.

Millipedes — Class Diplopoda

Like centipedes, the bodies of millipedes are generally elongated and have multiple segments. The segments are hardened and in cross-section form a complete ring in some species. In other species they form a dome, which allows the millipede to roll up into a nearly perfect ball. Each segment bears two pairs of jointed legs. The name "millipede" means 1000 legs. However, the largest number of legs any diplopod possesses is 375 pairs, for a total of 750 legs. They are generally short and appear to move in waves as they propel the millipede slowly forward.

Some species have a single ocellus (simple eye) on each side of the head, but some have 90 ocelli per side, while the soil dwelling species have none at all. Millipedes have one pair of short, elbow jointed antennae responsible for the senses of touch and smell, and are indispensible to the millipede's survival. Diplopods are vegetarians, feeding on living and dead plant material.

Approximately 8000 millipede species have been described by science. Some of the smallest, belonging to the genus Polynexus, are bristly creatures measuring only 3 mm (1/10"). By contrast, the largest members of the order Spirostreptida attain a length of 30 cm (11 3/4"). One of the Southwest's largest millipedes, *Orthoporus ornatus,* belongs to this order. Though diminutive by comparison, it grows to a substantial length of at least 12 cm (4 1/2"). Several other large species occur in the Southwest. *Orthoporus pontis,* a close relative, is a western Texas species. The more distantly related *Narceus americanus,* is a primarily southeastern species, but has a range which spills over into our region.

Millipedes occur most commonly in the temperate and tropical regions. Those species which inhabit arid regions have special physical and behavioral adaptations which make it possible for them to survive. Their thick, waxy exoskeletons inhibit water loss and they can go dormant during long periods of severe heat or drought.

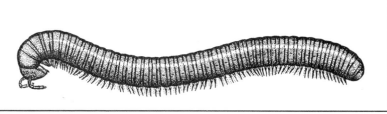

Desert millipede. Note the small pores (on the sides of the body) through which the millipede emits its caustic secretion when disturbed. *(Illustration by Jeffrey L. Martin)*

COMMON NAME: Desert Millipede

SCIENTIFIC NAME: *Orthoporus ornatus*

IDENTIFICATION: Long, cylindrical, segmented body. These millipedes can be as large as 12 cm (4 1/2") or more in length. They are light to dark brown in color and have two pairs of legs per body segment. Despite the number of legs, they move slowly. Males and females are similar in size and coloration.

RANGE: Sonoran and Chihuahuan deserts.

HABITAT: Under wood, rocks or decaying vegetation.

REPRODUCTION: Mating season begins with the onset of summer rains. The female stores sperm (from the male) until she is ready to lay her eggs. She lays up to 300 eggs, fertilizing them as they are laid in a nest of soil and fecal material. Several weeks later, the eggs hatch. Newly hatched larvae have only three pairs of legs. The number of legs and segments increases with each molt, and the cast skins are eaten. There are as many as 15 molts before adulthood. Desert millipedes live up to three years.

FOOD: Decaying and living vegetation.

DEFENSE: Desert millipedes have two glands per segment, one on each side, which produce irritating chemicals including quinones (highly volatile hydrocarbons). When disturbed, the millipede ejects droplets of this secretion through special pores to deter would-be predators, such as spiders and lizards. In humans, this secretion can cause blistering of the skin and considerable burning if rubbed in the eyes. Wash the afflicted skin area with soap and water. If the secretion gets in your eyes, flush with water until pain subsides and contact a physician.

Insects

Class Insecta

The insects are the most successful animals on earth. They have exploited every conceivable niche available. To date, over one million species have been described (more than all other animal species combined), with perhaps another million as yet undiscovered. Estimates of the global insect population range upwards to one thousand million billion. Their sensory systems and appendages are the most highly developed of all the arthropods. They are also the only arthropods that have evolved wings (millions of years ahead of any vertebrate), which allow them to disperse over wide areas. Relatively short life cycles and large broods allow them to evolve quickly in response to changes in their environment.

Although tremendous variation exists between species, members of this class have a basic anatomical format: three distinct body sections — the head, the thorax and the abdomen. Insects have three pairs of jointed legs and usually two pairs of wings. Their heads carry mouth parts, sensory antennae, and simple and compound eyes. The thorax is divided into three segments, each of which bears one pair of legs. The mid and hind segments bear one pair of wings each. The multisegmented abdomen contains breathing, reproductive and sometimes sensory apparatus.

Insect Reproduction

As a result of the stress of natural selection, insects have evolved a variety of reproductive strategies to ensure their survival. Internal fertilization and egg laying is the prevalent method, but the picture gets somewhat complicated after this. For example, wasps of the family Vespidae, which includes yellowjackets, produce viable, reproductive males from unfertilized eggs. The eggs are laid in time for the males to mature by the end of summer for the purpose of mating. Unfertilized female aphids (family Aphididae) on the other hand, do not lay eggs, but bear live *females* throughout the summer. As fall approaches, a generation of males and egg-laying females are born. This generation mates and eggs are laid to start the next season's process over again.

These examples are exceptions, rather than the rule. The rest of the insect world more or less conforms to the following means of reproduction:

1. Complete Metamorphosis

The vast majority of insects go through complete metamorphosis: egg, larva (caterpillar), pupa (rest stage), adult. Beetles, flies, bees and butterflies are among the insects which develop by means of complete metamorphosis.

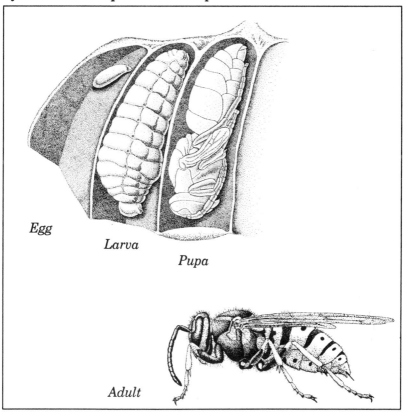

Complete metamorphosis of a **yellowjacket.** *(Illustration by Jeffrey L. Martin)*

Basically, the larva hatches, feeds and molts (sheds its skin) periodically, which allows it to grow. The period between molts is known as an instar. The final molt results in the pupal stage. In this stage, an array of hormonal signals initiates a breakdown and rearrangement of body material into new structures. When this process is complete, and when weather conditions are correct, the pupal shell is broken, and an adult insect emerges. In some species, only the larvae feed, and the adults emerge with mouthparts that are underdeveloped or lacking altogether. The role of the adult in these cases is reproduction only. Many moth species exhibit this characteristic. Their entire adult lifespan is a month or less. After mating and laying their eggs, they die.

This is not the case with all insects however. An excellent example is the ground beetles, which are voracious predators as adults as well as in the larval stage. As a rule, bees, ants, flies and other insects also feed throughout their lives.

Hypermetamorphosis

A variant of complete metamorphosis, called hypermetamorphosis, exists among certain insects. This large word merely means that from one instar to the next, the appearance and lifestyle of the larva is radically different. Blister beetles, which are covered in this book, are an excellent example of this phenomenon.

2. Incomplete Metamorphosis

Grasshoppers, cicadas, mantids and true bugs are among those insects which undergo incomplete metamorphosis: egg, nymph, adult. The nymph resembles a tiny, wingless adult upon hatching from its egg. It proceeds to feed and grow, molting until it reaches full size and sexual maturity.

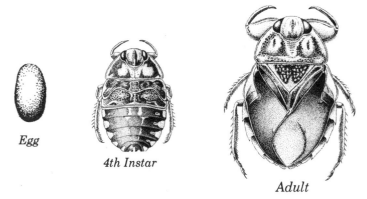

Egg

4th Instar

Adult

Giant Water Bugs go thru 5 instars before adulthood. *(Illustration by Jeffrey L. Martin)*

Bees, Wasps and Ants

(Order Hymenoptera)

This large order contains over 100,000 species worldwide. Most of these insects have two pairs of sparsely veined wings, but some are wingless. All have chewing mouthparts. Bees and wasps are also equipped with tonguelike structures for drinking nectar. Females possess an ovipositor, which is used for depositing eggs. In bees, wasps and ants, the ovipositor has evolved into a stinger. Most hymenopterans have a narrow waist between the thorax and abdomen (this feature is especially pronounced among the dauber wasps).

A number of species in this order are intriguing in that they are social animals, a rarity in the animal kingdom (and certainly among arthropods). The complexity of organization and specialization, as well as the co-operation among individuals, rivals even human society. There is a definite caste system in place (queen, worker, soldier), and many and various jobs are assigned to each caste. In all cases, these societies are primarily female, with a tiny minority of males in place for reproductive purposes only.

Most hymenopterans, such as carpenter bees, *Pepsis* wasps and velvet ants, are solitary. Many hunt crickets, spiders and caterpillars to feed their young. The adults themselves feed on nectar. Collectively, bees and wasps are prolific pollinators and a major factor in the evolutionary strategy and success of flowering plants.

The danger in the sting of these insects lies in an allergic

Comparison of stingers, left to right: paper wasp (smooth), honey bee (profusely barbed), and yellowjacket (slightly barbed). *(Photo by Garland McIlveen, Jr., Texas Agricultural Extension Service)*

reaction, which can be severe and even life threatening. More people die from the stings of ants, bees and wasps, than from the bites and stings of all other venomous animals combined. Allergic reactions usually (but not always) occur because the body has become sensitized by an earlier stinging incident. In effect, the body has armed itself in preparation for the next sting. But on occasion the response is too strong for the body to handle, and serious complications arise.

The **leafcutter bee** cuts neat circular pieces from leaves and flower petals to use for nesting material. *(Illustration by Jeffrey L. Martin)*

COMMON NAME: Leafcutter Bee

SCIENTIFIC NAME: Family Megachilidae

IDENTIFICATION: Leafcutter bees are from 10mm (7/16") to 14mm (9/16") in length. They are dark grey or black with whitish or golden hair on head, thorax, abdomen and legs. Their wings are clear or black. Female bees have specialized hairs on the underside of their abdomens, known as pollen brushes, that are used to collect pollen. The males are slightly smaller than the females and lack the pollen brush.

RANGE: Western United States.

HABITAT: Meadows, orchards, gardens and semiarid shrublands.

REPRODUCTION: The female builds a nest in holes in rotten wood or hollow plant stems or burrows into the soil (depending on species). She cuts a circular or oval piece from a leaf, rolls it up in her legs, flies back to the nest, and uses the leaf to line it. A ball of paste made from pollen and nectar is placed in the nest cell. An egg is then deposited on the pollen ball. Finally, the cell is sealed with a leaf fragment cut to fit the opening. The larva hatches and feeds on the pollen ball. It spends the winter as a pupa and emerges as an adult in the spring.

FOOD: Adults feed on nectar.

STING: Moderately painful but pain subsides quickly. Wash with soap and water, and apply an antiseptic. In the event of an allergic reaction, consult a physician.

Bumble bees are active during the spring and summer months, drinking nectar and pollinating flowers. *(Illustration by Jeffrey L. Martin)*

COMMON NAME: Sonoran Desert Bumble Bee

SCIENTIFIC NAME: *Bombus sonorus*

IDENTIFICATION: Adult bee body lengths vary according to caste: male drones and workers measure 9mm(3/8") to 14mm (1/2"); and the queen, 14mm (1/2") to 18mm (3/4"). They have very sturdy bodies, with a velvety covering of black and yellow hair. The wings are smoky-colored.

RANGE: Southwestern deserts.

HABITAT: Desert environments of flowering plants.

REPRODUCTION: Sonoran Desert bumble bees are social bees and live in colonies. Unlike honeybees, the colony does not survive the winter. Only mated queens overwinter (hibernate), emerging in early spring to begin a new colony. A queen locates a rodent's abandoned burrow suitable to serve as a brood chamber. She makes small, round wax brood cells in which she deposits pollen, honey and 7 to 16 eggs. She also makes cells called honeypots, which are intended for nectar storage. The larvae grow and pupate, emerging as females (workers). A mature nest may have 100 or more individuals, but the population seldom exceeds 400 bees. In late summer, reproductive females and drones are produced. They fly off to mate with bees from other nests.

FOOD: Adults drink nectar.

STING: The bumble bee can sting repeatedly. The site of the sting becomes a dark red spot surrounded by a swollen red area. Swelling usually lasts no more than two days. Wash with soap and water, and apply an antiseptic. In the event of multiple stingings, or an allergic reaction, consult a physician.

NOTE: Bumble bees are defensive in the proximity of their nest and rigorously pursue any creature unfortunate enough to stumble upon it. However, most encounters are accidental, and stinging incidents are uncommon. Away from the nest, bumble bees are relatively unconcerned and tolerant.

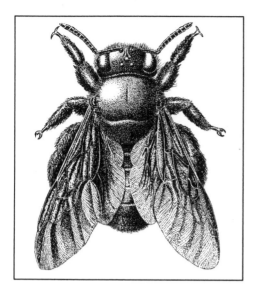

The female **carpenter bee,** pictured here, can sting but is not very aggressive. Males cannot sting but may be aggressive to nesting site intruders. *(Illustration by Jeffrey L. Martin)*

COMMON NAME: Carpenter Bee

SCIENTIFIC NAME: *Xylocopa* sp.

IDENTIFICATION: Adult carpenter bees are 20mm (3/4" to 25mm (1") in length. Females are glossy blue-black in color and resemble bumble bees in shape. Males are golden colored.

RANGE: Utah, Nevada, California, Oregon, Arizona to Mexico.

HABITAT: Forests, meadows, deserts.

REPRODUCTION: The female chews a tunnel nest in dead trees, agave stalks, softwood lumber, etc. She then creates chambers within the tunnel by forming walls with plant material or sawdust mixed with her saliva. Two or more pollen balls and one egg are deposited in each chamber. The larva feeds on the pollen ball and pupates. Adults emerge in late summer.

FOOD: Adults drink nectar.

STING: Stings are painful but only for a few minutes. Wash the sting site with soap and water, and apply an antiseptic. Consult a physician if severe or unusual symptoms occur.

Digger bees are neither aggressive or defensive of their nests, but will sting if restrained. *(Illustration by Jeffrey L. Martin)*

COMMON NAME: Digger Bee

SCIENTIFIC NAME: *Centris* sp.

IDENTIFICATION: Digger bees range from 15mm (9/16") to 17mm 5/8") in length. They have short, dense hair on their head, thorax and legs. The wings are clear.

RANGE: Most of North America.

HABITAT: Meadows, fields and gardens.

REPRODUCTION: Females nest in soil. They line their burrow with a thin layer of wax as a support against crumbling soil. Each cell receives one egg and is provisioned by the female with mixtures of honey and pollen. The larvae feed and pupate during the winter and emerge as adults in late spring. Adults also spend the winter in burrows. Some males dig virgin females out of burrows and mate; others find mates near flowers where females are likely to visit.

FOOD: Adults drink nectar.

STING: The digger bee sting is very painful, but effects are local to the sting. Wash the sting site with soap and water, and apply an antiseptic. Ice may be used to ease pain.

The **honey bee** feeds on nectar. It collects pollen on bristly hairs on its hind legs called the pollen basket. *(Photo by Dr. Gerald Loper)*

COMMON NAME: Honey Bee

SCIENTIFIC NAME: *Apis mellifera*

IDENTIFICATION: Workers (non-reproductive females) range from 15mm (9/16") to 17 mm (5/8"). Drones (male), 18mm (11/16") to 24mm (15/16"). Queens (reproductive females), 24mm 15/16") to 30mm (1 3/16"). These bees are reddish-brown with black rings on their abdomens. Wings are translucent. The hair on their body and legs makes them look fuzzy.

RANGE: Worldwide. Introduced into the United States by European immigrants for domestic honey production. Escaped bees set up wild colonies.

HABITAT: Meadows, open woods, gardens, hollow trees and in beekeepers' hives.

REPRODUCTION: Queen bees may lay as many as 2000 eggs daily. Workers (non-reproductive females) tend to the eggs and larvae as they develop in the honeycomb. The adult lifespan of a worker bee is approximately three months while queens live from two to four years. Drones (males) are maintained for mating only and perform no other function in the hive. Mating occurs, (usually in flight) and the male then dies.

Hive numbers can be in the tens of thousands. If a queen is needed (old queen dies or weakens, hive is overpopulated and needs a new queen to start a new hive, etc.) workers feed a special nutritive substance called royal jelly to two or three of the larvae that have been kept in specially enlarged cells. These larvae are fed up to 50% more than the worker larvae so they grow faster and are much larger in size. Upon emergence, the new queens fight to determine which is the strongest. The weaker queens are killed and discarded.

Worker bees eat pollen and nectar as they forage for food for the hive, a little for the worker, a little for the hive. They also bring water, which is used to dilute honey and to create an evaporative cooling system for the hive.

FOOD: Nectar and pollen for workers, drones and most larvae. Queen larvae are fed royal jelly. This is a thick substance secreted chiefly by two glands in the worker bee's head. These glands have openings to the worker's mouth. All larvae are fed royal jelly for the first three days of their life. After that, only

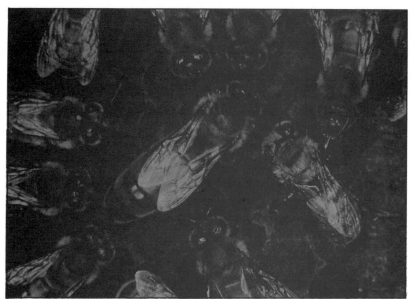

Honey bee workers constantly tend to the queen (center), feeding her royal jelly and caring for her larvae. *(Photo by Dr. Gerald Loper)*

the queen larvae continue on this special food.

STING: Honey bees will defend their hive when it is disturbed, and do not hesitate to sting. Because its stinger is barbed, once inserted it is pulled out of the bee's body and the bee dies. The stinger works its way deep into the skin and a poison sac pumps venom into the wound. The pain from this sting is not long lasting. Scrape the stinger out of the skin. Do not pinch it as this will squeeze more venom into the wound. Wash with soap and water, and apply an antiseptic. In cases of allergic reaction including anaphylactic shock, consult a physician. *(See First Aid Section.)*

NOTE: Do not approach *any swarm* of bees unless you are properly clothed, equipped and trained. If you think you have seen Africanized honey bees, or see a bee swarm of any kind in your neighborhood, report the sighting to your local fire department and department of agriculture.

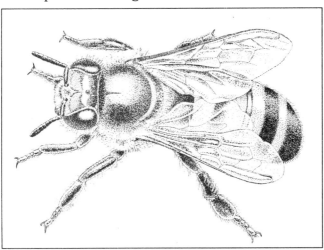

The **Africanized honey bee** or "killer bee" is spreading into the Southwest from South America. *(Illustration by Jeffry L. Martin)*

COMMON NAME: Africanized Honey Bee (also known as "killer bees")

SCIENTIFIC NAME: *Apis mellifera scutellata*

IDENTIFICATION: It is impossible to distinguish this bee from the honey bee without a microscope (leg hair, wing veins, etc.) or by specialized protein analysis in a laboratory.

RANGE: These bees were imported from South Africa to Brazil in 1956. Wild colonies then spread from Argentina to Panama and continue northward. They were first identified in U.S. in the mid-1980's and may be found in the southern parts of California, Arizona, New Mexico and Texas.

HABITAT: Meadows, deserts, woods. As their requirements are the same as honey bees, they inhabit similar environments.

REPRODUCTION: Identical to honey bees.

FOOD: Nectar and pollen.

STING: The Africanized honey bee is more aggressive than the common honey bee and usually attacks in larger numbers. Because of this, such an attack could be fatal. First aid for stings includes washing the area with warm water and soap, and applying an antibiotic. In cases of multiple stings or allergic reaction including anaphylactic shock, consult a physician. *(See First Aid Section.)*

The **velvet ant** is actually a species of wasp. Females are wingless. *(Illustration by Jeffrey L. Martin, courtesy Arizona-Sonora Desert Museum)*

COMMON NAME: Velvet Ant

SCIENTIFIC NAME: Family Mutillidae

IDENTIFICATION: The velvet ant is actually a primitive wasp. Adults range in size from 5mm (5/16") to 25 mm (1"), depending on species. Female velvet ants are wingless. Males are larger than the females and have wings. Both have dense medium to long hairs on body and legs. Colors are monochromatic or can be a combination of red, orange, yellow, white or black.

RANGE: Riparian and desert areas in the southwest.

HABITAT: Arid open desert, washes and open fields.

REPRODUCTION: Female velvet ants search for burrows of ground-nesting bees and wasps. They then lay their eggs on the larvae or pupae of the host. The larvae feed on the host until ready to pupate. Adults emerge in the spring.

FOOD: Adults drink nectar.

STING: This wasp produces a scraping, squeeking noise from its abdomen when provoked. It has a very painful sting, but pain usually subsides within an hour. Wash the site with soap and water, and apply an antiseptic. In case of infection or unusual symptoms, consult a physician.

Bald-faced hornet. Unlike honey bees, the hornet's stinger is not barbed. As a result, hornets can sting repeatedly. *(Photo by Garland McIlveen, Jr., Texas Agricultural Extension Service)*

Large paper nest of the **bald-faced hornet.** A single nest may contain up to 10,000 hornets. They will defend it with the slightest provocation. (*Photo by Garland McIlveen Jr., Texas Agricultural Extension Service*)

COMMON NAME: Bald-faced Hornet

SCIENTIFIC NAME: *Vespula maculata*

IDENTIFICATION: 16mm (5/8") to 20mm (3/4"). Bald-faced hornets are black overall with bone-white markings on face, thorax, and abdomen. Their wings are smoky-colored.

RANGE: Throughout North America.

HABITAT: Forests, meadows, suburban areas.

REPRODUCTION: In the spring the queen bald-faced hornet chews wood to build a small pendant nest out of gray pulp. In this nest she lays eggs and produces a first generation of only female workers. It is this generation and later generations of workers that build and maintain the nest. In the fall, after producing a new generation of males and queens, the old queen, her workers, and any existing young all die. Only new young mated females overwinter in the soil or among ground litter.

FOOD: Adults drink nectar, and eat small insects.

STING: Baldfaced hornets aggressively defend their nests and will actively pursue any creature unfortunate enough to disturb them. An individual wasp can sting repeatedly. The extremely painful sting delivers a complex venom containing neurotoxins and other components capable of causing tissue destruction local to the site of the sting. Do not apply ice to wound. Wash with soap and water, and apply an antiseptic. A paste of baking soda can be applied to relieve pain. In the event of a multiple stinging incident or allergic reaction including anaphylactic shock, consult a physician.

Southern yellowjackets *(Vespula squamosa)* entering and exiting their nest. *(Photo by Garland McIlveen, Jr. Texas Agricultural Extension Service)*

COMMON NAME: Yellowjacket

SCIENTIFIC NAME: *Vespula & Dolichovespula* sp.

IDENTIFICATION: Adult workers range from 10mm (3/8") to 12mm (1/2"), queens 12mm (1/2") to 16mm (5/8"), depending on the species. Yellowjackets have striking yellow and black markings on head, thorax and abdomen. Their wings are smoky-colored.

RANGE: Throughout the United States.

HABITAT: Meadows, edges of forested land, stumps and fallen logs. Yellowjackets are frequently found in urban areas.

REPRODUCTION: Yellowjacket nests are of paper-like material constructed either underground or in trees depending on the species. The mated, female yellowjackets spend the winter under loose bark, in cracks or in rock crevices. In the spring

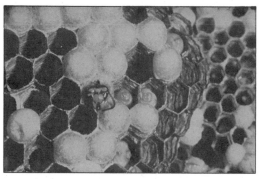

Southern yellowjackets (*Vespula squamosa*). Developing brood in comb cells. *(Photo by Garland McIlveen, Jr., Texas Agricultural Extension Service)*

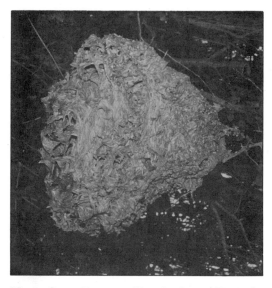

Nest of **southern yellowjackets** *(Vespula squamosa)* attached to a tree branch. *(Photo by Garland McIlveen, Jr., Texas Agricultural Extension Service)*

Scorpions
and **Venomous**
Insects
of the
Southwest

Note: Even though color photographs assist us in identifying these denizens of the Southwest, some of them are still difficult to see, as they have been photographed in their natural environment where the ability to blend in with their surroundings is an important factor for survival.

Bark, Devil & Giant Desert Hairy Scorpions (p. 10)

Striped Centruroides stinger (p. 12)

Striped Centruroides (p. 13)

Devil Scorpion (p. 13)

Giant Desert Hairy Scorpion
(p. 14)

Windscorpion
(p. 17)

Whipscorpion
(p. 18)

Whipscorpion
(p.19)

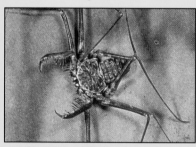

Tailless Whipscorpion
(p. 21)

Tarantula
(p. 25)

Tarantula
(p. 25)

Wolf Spider
(p. 28)

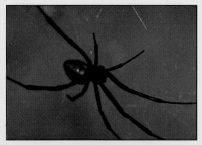

Black Widow Spider
(p. 29)

Black Widow Spider
(with egg sac)   (p. 29)

Brown Spider
(p. 31)

**Severe scarring caused by
Brown Spider bite.**   (p. 31)

Crab Spider
(p. 33)

Crab Spider (eating Leaf
Beetle)   (p. 34)

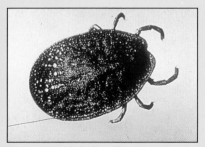

Adobe Tick
(p. 36)

Rocky Mountain Wood Tick
(p. 37)

Brown Dog Tick
(l) female *(r)* male (p. 38)

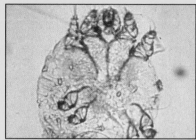

Scabies Mite
(p. 39)

Giant Desert Centipede
(p. 43)

Giant Desert Centipede
(underside of head) (p. 43)

Desert Millipede
(p. 47)

**Paper Wasp, Honey Bee &
Yellowjacket Stingers** (p. 51)

(l-r) **Leafcutter, Honey &
Bumble Bees** (p. 52)

Bumble Bee
(p. 53)

Honey Bee
(p. 57)

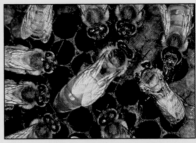

Honey Bee workers & queen
(p. 57)

Africanized Honey Bee
(killer bee) (p. 59)

Velvet Ant
(p. 60)

Bald-faced Hornet
(p. 62)

Bald-faced Hornet Nest
(p. 62)

Yellowjacket
(p. 63)

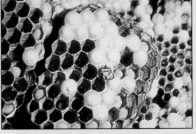

Southern Yellowjackets
(in comb cells) (p. 64)

Southern Yellowjacket Nest
(p. 64)

Southern Yellowjackets
(p. 64)

Paper Wasp
(p. 65)

**Paper Wasp foundresses
at work** (p. 65)

Paper Wasp
(p. 65)

Paper Wasps
(p. 65)

Cicada Killer (with prey)
(p. 67)

Cicada Killer (mounted)
(p. 67)

Blue Mud Dauber
(p. 70)

Mud Dauber nest
(p. 70)

Blister Beetle
(p. 76)

Darkling Beetle
(p. 77)

Conenose nymph
(p. 80)

Giant Water Bug
(p. 82)

Adult Mosquito
(p. 84)

Larval Mosquito
(p. 86)

Mosquito Pupae
(p. 86)

Human Flea
(p. 89)

Head Lice
(l) female *(r)* male (p. 91)

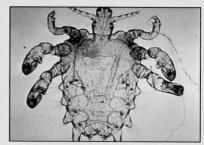

Crab Louse
(p. 94)

Puss Moth Caterpillar
(p. 96)

Buck Moth Caterpillar
(p. 97)

Color photos courtesy of:

Arizona-Sonora Desert Museum
J. W. Stewart
Teaching Files, Dept. Entomology, Texas A & M University
Entomological Society of America
Terry Christopher
Francis Brady
Jeffrey L. Martin
Garland McIlveen Jr., Texas Agricultural Extension Service.

they build paper comb cells and lay a single fertilized egg in each cell. Nests may be built underground or suspended from a tree branch, and may grow large enough to raise in excess of 12,000 individuals. Females feed the larvae insects and carrion of any kind plus BBQ chicken, hot dogs, and hamburger. Larvae pupate and emerge as worker adults throughout the spring and summer season. Males and reproductive females are produced in the fall. They mate with individuals from other nests. Only mated females survive the winter.

FOOD: Adult yellowjackets drink nectar and juices from rotting fruit, scavenge animal carcasses and hunt insects to feed larvae.

STING: Yellowjackets will readily defend their nest. Because the stinger is not barbed, an individual wasp can sting repeatedly. The painful sting delivers a complex venom containing neurotoxins and other components, which are capable of causing tissue destruction local to the site of the sting. Do not apply ice to wound. Wash with soap and water, and apply antiseptic. A paste of baking soda can be applied to relieve pain.

Worker (female) **paper wasps** share nest-making responsibilities and care of the larvae. *(Photo by Garland McIlveen, Jr., Texas Agricultural Extension Service)*

COMMON NAME: Paper Wasp or Umbrella Wasp

SCIENTIFIC NAME: *Polistes* spp.

IDENTIFICATION: These wasps average 25mm (1") in length. Depending on their species, they are yellow or brown overall, or a combination of these colors. Some have black markings as well. They have a short, somewhat thick waist (pedicel). Their wings are smoky to black.

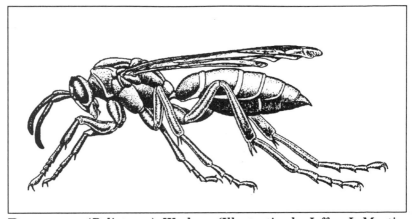

Paper wasp *(Polistes* sp). Worker. *(Illustration by Jeffrey L. Martin, courtesy of Arizona-Sonora Desert Museum)*

RANGE: Throughout the United States.

HABITAT: Meadows, fields, deserts.

REPRODUCTION: In the spring one or more females (called foundresses) build a hanging, umbrella-shaped paper nest from a mixture of wood pulp and saliva. One female becomes the dominant queen of the nest and she usually does all of the egg laying. These nests (uncovered tiers of cells) can frequently be found under the eaves of buildings. The nests hang from a thin stalk (petiole) and are open at bottom to allow the adults to feed the larvae prechewed insects (such as butterflies and moth larvae). The larvae pupate and emerge as worker (females) adults throughout the spring and summer season. Unfertilized eggs produce reproductive males and fertilized eggs, females. They mate with wasps from other nests. Only mated females survive the winter, sheltered in leaf litter and rock crevices, etc. Males, unmated females, the young and the old queen die.

FOOD: Adults drink nectar and juices from rotting fruit.

STING: Paper wasps will readily defend their nest. Because the stinger is not barbed, an individual wasp can sting repeatedly. The painful sting delivers a complex venom capable of causing tissue destruction local to the site of the sting. First aid for stings includes washing the area with warm water and soap and applying an antibiotic. In cases of multiple stings or allergic reaction including anaphylactic shock, consult a physician. *(See First Aid Section.)*

Cicada killer with prey. *(Photo courtesy of Teaching Files, Dept. Entomology, Texas A&M University)*

COMMON NAME: Cicada Killer

SCIENTIFIC NAME: *Sphecius speciosus*

IDENTIFICATION: Female cicada killers are large, achieving up to 50mm (2") in length. Their head and thorax are brown. Abdomens are black with yellow or cream-colored markings, usually three pairs, on each side. Wings are transparent. Males and females are similar in color, but the male is the smaller of the two.

RANGE: Throughout North America.

HABITAT: Forests, deserts, wherever cicadas are found.

REPRODUCTION: Adults are usually seen from June through August in the Southwest, coinciding with the emergence of

Mounted **cicada killer**. Note stinger and transparent wings. *(Photo by Garland McIlveen, Jr., Texas Agricultural Extension Service.*

cicadas. Mating occurs on the ground, but the coupled mating pair can fly if necessary. If a female does not mate, her eggs will bear male offspring; if she mates, her offspring will all be female. The female tunnels into the soil and creates branches of two or more brood cells, each of which is large enough to accept one or two cicadas. She then hunts for cicadas, which she paralyzes with her sting, and places in the brood cells (one cicada for a male larva, two for a female). One egg is deposited per cell. Larvae feed on the cicadas and pupate in the cell. Adults emerge the following spring.

FOOD: Adults drink nectar and feed cicadas to their larvae.

STING: Nonaggressive. Stings only if restrained. Sting is not serious for humans, and pain is short lived. Wash sting site with soap and water, and apply disinfectant. Consult your physician in the event of unusual symptoms.

COMMON NAME: Pepsis Wasp, Tarantula Hawk

SCIENTIFIC NAME: *Pepsis, Hemipepsis* spp.

IDENTIFICATION: Pepsis wasp adults are approximately 33 mm (15/16") or more in length. Head, thorax, abdomen and legs are glossy blue-black. Wings are reddish to orange. One species, *H. pattoni*, has blue-black wings. Both sexes are similar in color.

RANGE: Southwestern United States.

HABITAT: Open desert, arid grassland, and lower mountain slopes.

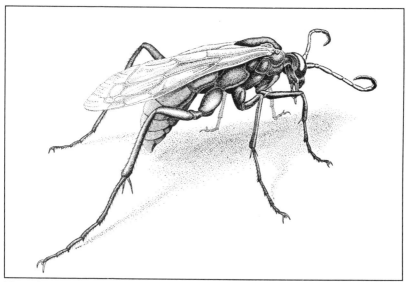

Pepsis wasps are called **tarantula hawks** because they hunt for tarantulas to feed their young. *(Illustration by Jeffrey L. Martin)*

REPRODUCTION: The female pepsis wasp immobilizes tarantulas or trap-door spiders with a paralyzing sting and then maneuvers them into an underground chamber. One large, white egg is deposited on the spider. The larva hatches within five days. It feeds on the spider, eating vital organs last so as to keep it alive until the larva is ready to pupate (about one month). The larva then spins a cocoon and spends the winter as a pupa. Adults emerge in the spring.

FOOD: Adults drink nectar.

STING: Nonaggressive. Sting is painful but only stings if restrained. Incidence of stingings is rare; therefore, severity of reactions is not well documented. Wash with soap and water, and apply an antiseptic. Consult a physician in the event of unusual symptoms.

NOTE: One species of robber fly (family Asilidae) is a pepsis wasp mimic. This slender fly is a predaceous, but harmless species.

Mud dauber *(Chalybion californicum)* on nest. *(Photo by Garland McIlveen, Jr., Texas Agricultural Extension Service)*

COMMON NAME: Mud Dauber

SCIENTIFIC NAME: *Sceliphron, Chalybion*

IDENTIFICATION: Up to 18 mm (11/16") in length. Distinctive long, slender waist. Iridescent blue-black. Some species have yellow areas on thorax, abdomen and legs.

RANGE: Throughout North America.

HABITAT: Daubers live on cliff sides and in meadows, usually near a source of mud for nesting. They build their nests under protective overhangs, on a vertical surface. Nests can be found under the eaves of houses and other buildings.

REPRODUCTION: The female builds a mud nest of tubular cells. She places in each cell one or more spiders she has paralyzed with a sting. She then deposits one egg (per cell) on the spider and seals the nest with mud. Larvae feed on the spiders, pupate, and emerge as adults.

FOOD: Adults drink nectar.

STING: Nonaggressive. Sting is relatively mild. Wash with soap and water, and disinfect. Consult physician in the event of an unusual reaction.

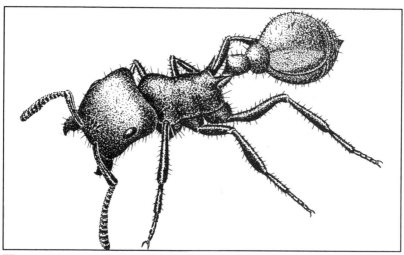

Harvester ant worker. Note large head, thin waist and spines at posterior end of the thorax. *(Illustration by Jeffrey L. Martin)*

COMMON NAME: Harvester Ant

SCIENTIFIC NAME: *Pogonomyrmex rugosus*

IDENTIFICATION: Workers (females) range from 6mm to 7mm (approximately 1/4") in length. The head and thorax of this ant are black, the abdomen reddish. Their heads are large, with characteristic long hairs on the underside. The head of the male is much smaller relative to body size than that of the female. The queen is much larger than all other colony members.

RANGE: Southwestern United States.

HABITAT: Cultivated fields, sandy or barren areas.

REPRODUCTION: The nest, which may contain 20,000 individuals, is built underground. The entrance may be a crater, crescent mound, or merely a flat disk-shaped area. The queen founding the new colony rears the first brood. First generation female workers then care for eggs and young, and feed the queen with regurgitated food. The male population in the nest is small until the summer rainy season when winged reproductive ants emerge from the nests to mate during swarming events.

FOOD: Harvester ants eat and store seeds and grains. They use an antibiotic saliva with which they coat their stored food to retard spoilage. Workers secrete a chemical substance from the tip of their abdomen (touching their stinger to the ground) to lay a scent trail between a food source and the nest.

STING: The harvester ant stings by clinging to the victim's skin with its mandibles and thrusting upward with its stinger. In humans, the sting is extremely painful. The site of the sting is marked by an inflamed red bump, which is followed by a small watery blister. Pain may last for as long as six hours. First aid includes washing the area with warm water and soap and applying an antibiotic. In cases of multiple stings or allergic reaction including anaphylactic shock consult a physician. *(See First Aid Section.)*

NOTE: Harvester ant trails are occasionally discovered leading into a house, or to a shed where grains are stored. Sealing doors and windows and caulking cracks are a good infiltration defense. A variety of ant traps and poisons are available. Read labels carefully and always consider the safety of pets and children when using them. Consult an exterminating service.

Southern fire ant. Fire ants usually cling to their victim with their mandibles and sting repeatedly. *(Illustration by Jeffrey L. Martin)*

COMMON NAME: Southern Fire Ant

SCIENTIFIC NAME: *Solenopsis xyloni*

IDENTIFICATION: These ants vary in size from 1.6 (1/16") to 6.6mm (1/4"). They have short hairs on their head, thorax and abdomen, and two pronounced "bumps" on their petiole (waist). Females have light reddish-brown heads. Thorax and abdomen are black. Males are reddish black with reddish yellow legs.

RANGE: Southern United States.

HABITAT: Capable of inhabiting a variety of desert habitats, from streams and washes to arid open desert.

REPRODUCTION: Nests in the ground in shaded areas, preferably in relatively damp soil. Nests can also be found in dead wood. Suitable artificial nesting sites include the space within walls of buildings. Fire ants are extremely protective of their nests and will aggressively attack an intruder.

FOOD: Seeds, flowers, fruit and plant juices, small insects, as well as human and pet foods. Secretes a chemical substance from tip of abdomen (touching stinger to the ground) to lay a scent trail between food source and nest. As they travel back and forth, the ants intermittently renew the trail.

STING: The southern fire ant stings by clinging to the victim with its mandibles and repeatedly inserting its stinger. The sting is very painful. A small bump develops, and within 24 hours a watery blister forms which itches intensely. Blisters rupture due to scratching; the sore scabs and slowly heals. Wash the area with warm water and soap and apply an antibiotic. In cases of multiple stings or allergic reaction, consult a physician. *(See First Aid Section.)*

COMMON NAME: Field Ant, Red Ant

SCIENTIFIC NAME: *Formica* spp.

IDENTIFICATION: These ants are red, brown, black or a combination of all three, depending on their species. They range from 3mm (1/8") to 12mm (1/2") long. Males and females are very similar in size and coloration.

RANGE: Throughout United States.

HABITAT: Wooded slopes at high elevations, desert floors, and urban areas.

REPRODUCTION: Field ants nest underground or in mounds of leaves and sticks around shrubs. In urban areas they also nest between cracks in the pavement. Mating occurs during

Field ant. Note single segment of petiole (waist) *(Illustration by Jeffrey L. Martin)*

swarming events triggered by environmental conditions (temperature, humidity). Swarming males and queens have wings, which they lose upon finding a mate. The male dies soon after mating. The queen tends the first brood of a colony. This brood of female workers then takes over care for eggs and young and feeds the queen with regurgitated food.

FOOD: The field ant raids the nests of other insects such as termites and ants and eats them. They also feed on nectar. They engage in a symbiotic relationship with aphids. Aphids secrete a sweet substance known as "honeydew" from a pair of thin tubes (called cornicles) projecting up from their abdomen. Ants tend the aphids like cattle and "milk" them by brushing the aphid's back with their antennae. This induces the aphids to secrete the honeydew, which the ants consume. The proximity of the ants affords the aphids protection from insect predators. As do many others, this ant secretes a chemical substance from the tip of its abdomen to lay a scent trail (which it renews occasionally) between food sources and nests.

BITE: Field ants spray formic acid (an irritating chemical) from glands in their anal region, into a bite-inflicted wound. This is strictly a defense mechanism and is not used to subdue the insects on which the ant feeds. Field ant bites are painful but not toxic. Wash with soap and water.

Beetles

(Order Coleoptera)

Coleoptera is the largest order in the animal kingdom, containing over 300,000 described species. These insects are characterized by their elytra (wing covers), which are actually hardened forewings. Their thick tank-like exoskeletons have adapted to many environments, and as an order these insects have evolved to fill every conceivable niche.

Many beetles, such as ground beetles (family Carabidae), are predatory, as are their highly mobile larvae. Scavengers, such as carrion and burying beetles (family Silphidae), feed on the bodies of dead animals. Weevils (family Curculionidae) have long slender snouts with which they drill into seeds, fruits, nuts and stems to feed. Diving beetles (family Dytiscidae) have taken to a predaceous life in the water. They are armored, turtle-like submarines, with legs modified into efficient swimming paddles. The females of bee parasite beetles (family Stylopidae) spend their entire lives protruding from between the abdominal segments of bees and wasps. Lacking appendages and sense organs, they have degenerated to breeding receptacles. Only the males are fully developed and mobile.

Coleopterans display a virtually endless variety of shapes, colors and patterns. They range in size from tiny dermestids (family Dermestidae), up to the giant longhorns (family Cerambycidae) and goliath beetles (family Scarabidae) of the tropics. Some beetles, such as fireflies (family Lampyridae), produce their own light, which they use to recognize each other for mating purposes. This is but a sampling of the diversity and adaptability that has made coleopterans so successful as a group.

Beetles undergo complete metamorphosis (egg, larvae, pupa, adult). Some larger beetles are capable of biting but are not venomous. Three beetle families (Carabidae: ground beetles, Tenebrionidae: darkling beetles, Meloidae: blister beetles) have developed chemical defense mechanisms, particularly secretions of varying degrees of offensiveness.

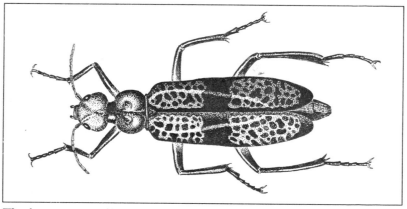

The **iron cross blister beetle** *(Tegrodera aloga)* gets its name from the black cross-shaped marking on its wing covers, which are otherwise yellow with black spots. *(Illustration by Jeffrey L. Martin)*

COMMON NAME: Blister Beetle

SCIENTIFIC NAME: Family Meloidae.

IDENTIFICATION: Blister beetles have soft, elongated bodies and elytra, 15mm (5/8") to 28mm, (1 1/8") in length. They have a visible neck between head and thorax. In some species the head is wider than the thorax. These beetles are usually gray or black, but some species exhibit warning colors (red, orange, or yellow) which announce toxicity.

RANGE: Southwestern deserts.

HABITAT: Deserts.

REPRODUCTION: Adults emerge in the spring and mate. They may be conspicuously numerous at this time. The female lays her eggs in soil. The larvae undergo hypermetamorphosis (take a different form with each molt). Newly hatched larvae (known as triungulins) are slender with long legs and are extremely mobile. With each molt, they become fatter, more grub-like and ultimately immobile. They then overwinter as a pupa. In some species, the triunguline climbs into flowers and onto the backs of solitary and social bees. They are unwittingly carried to that bee's nest, where they feed on its pollen and honey. They also eat the bee's eggs and larvae. Larvae of other blister beetles forage underground for grasshopper eggs.

FOOD: Adults eat plant tissue (flowers, leaves, etc.). Some species are agricultural pests.

DEFENSE: Blister beetles can pressurize their bodies and exude hemolymph (insect blood) from their bodies which contains the chemical cantharidin. At first tingling and rash occurs on the skin and then, within hours, blistering. Wash the area with soap and water, apply antiseptic and bandage. Care should be taken to prevent blisters from breaking. If blisters break, wash again with soap and water; reapply antiseptic and bandage. Consult a physician in the event of a severe condition.

Darkling beetles have fused elytra (wing shells) and are therefore flightless. *(Photo by Jeffrey L. Martin)*

COMMON NAME: Darkling Beetle

SCIENTIFIC NAME: Family Tenebrionidae

IDENTIFICATION: These beetles range in length up to 32mm (1 1/4"), depending on the species. They are dull brown, grey or black in color with fused elytra (forewings). Their bodies are moderately robust, torpedo-shaped and very hard. Darklings have long stout legs and walk with their heads down.

RANGE: Southwestern deserts.

HABITAT: Under boards, stones, fallen wood and debris. Active during the night or on cloudy days.

REPRODUCTION: Mating occurs during the summer. A pair will mate repeatedly. Male darkling beetles ride on the backs of females during and between mating. Females lay their eggs in moist soil. Larvae feed on dead plant and animal material.

FOOD: Leaf litter of desert shrubs.

DEFENSE: Darkling beetles of the genus *Eleodes* have a chemical defense mechanism. When disturbed, this beetle will raise its abdomen high in the air, as a warning signal. If aggravated further, an odorous golden brown liquid is released from the tip of the abdomen. This liquid effectively protects the beetle from a variety of predators. The strong acrid odor pervades the air for at least 15 minutes after the beetle is disturbed.

This same liquid on human skin causes tan-colored spots that cannot be washed off but will eventually wear off. The affected area should be washed with soap and water. Avoid contact with nose, eyes and mouth, as severe pain and inflammation can result. Flush these areas with water.

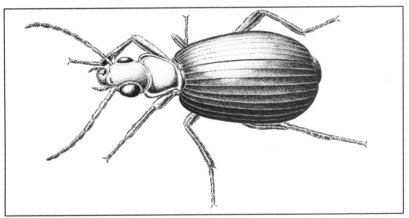

The **bombardier beetle** is capable of firing repeatedly. Would-be aggressors are greeted with a faceful of irritating, odorous vapor. *(Illustration by Jeffrey L. Martin)*

COMMON NAME: Bombardier Beetle

SCIENTIFIC NAME: *Brachinus* spp.

IDENTIFICATION: Bombardier beetles range in length from 5mm (1/4") to 15mm (5/8"). Their head, thorax, antennae and legs are brownish yellow or orange. The elytra (forewings) are blue or black and have ridges that run lengthwise. Males and females are similar in size and coloration.

RANGE: Throughout United States.

HABITAT: Margins of streams, lakes, and ponds. This is a nocturnal scavenger that hides under leaves, rocks and logs during the day.

REPRODUCTION: Mating takes place at night. The female lays eggs in cells in moist sandy soil. The larvae of this beetle are parasites of the larvae of aquatic whirligig beetles. The newly hatched bombardier larva finds and clings to a full-term whirligig larva leaving the water to build its mud cocoon. The bombardier clings to this host, feeds on it and uses its cocoon to pupate over the winter. Adults emerge during spring rains.

FOOD: Adults scavenge dead vertebrates and invertebrates, such as frogs and insects.

DEFENSE: This beetle derives its common name from the vaporizing liquid which it emits as a defense against predators. The bombardier beetle has two reservoirs in its abdomen, where hydroquinones secreted by special glands are stored. When the beetle is disturbed, these chemicals are dumped into two reaction chambers, where hydrogen peroxide is instantaneously secreted. The resulting reaction is a hot (212° F - 100°C) stinging spray from the tip of the abdomen. The beetle can accurately direct this spray at would be predators.

In humans, this spray can cause skin discoloration, skin and eye irritation, and inflammation of mucous membranes. Wash affected skin areas with soap and water. Flush eyes, nose and mouth with water if contaminated.

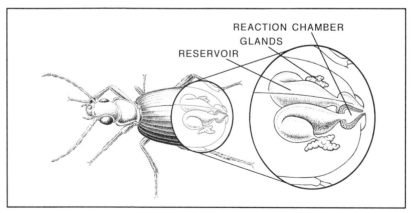

The organs of the bombardier beetle's elaborate defense mechanism are located in the rear of its abdomen. *(Illustration by Jeffrey L. Martin)*

True Bugs
(Order Hemiptera)

Members of this order are characterized by the forewings, which are hardened and leathery near their base, and membranous at their ends (hence "half-wing," the literal meaning of "Hemiptera"). Most true bugs have well developed wings.

The true bugs all possess jointed, beak-like mouth parts which they use for piercing and sucking. Some feed on plants. Others are predaceous. Some hemipterans are parasitic, feeding on the blood of mammals.

A number of species have taken up life in freshwater. Some are crawlers, but many are active swimmers. Aquatic hemipterans are largely predaceous, but the water boatmen (family Corixidae) also feed on algae, and microscopic plants and animals.

True bugs undergo incomplete metamorphosis (see illustration, page 49). From the moment they hatch, the young are miniature wingless versions of their parents. The young are known as nymphs. In nearly all species, the nymphs go through five instars. Only after their fifth and final molt are their wings fully developed and functional.

This *Triatoma infestans* nymph has yet to grow its wings. *(Photo courtesy of Teaching Files, Dept. Entomology, Texas A&M University)*

COMMON NAME: Conenose Bug

SCIENTIFIC NAME: *Triatoma* sp.

IDENTIFICATION: Conenose bugs range in length from 14mm (9/16") to 24 mm (1"). They are dark brown to black with

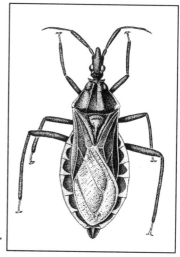

Adult **conenose bug** (*Triatoma rubida*). Note markings on the edges of the abdomen and the cone-shaped "nose". *(Illustration by Jeffrey L. Martin)*

red, orange, or yellow markings on the thorax, abdomen and at the base of the wing margin. The name derives from the conical head from which the straight beak protrudes.

RANGE: Coastal and desert United States, Central and South America.

HABITAT: Southwestern deserts. These insects are attracted to porch lights and windows at night. They seek shelter from the sun during the day, entering homes through cracks around windows and under doors.

REPRODUCTION: Mating takes place during the spring and summer. The female lays her eggs in the burrows of small rodents. The eggs hatch in 3 to 5 weeks. All five molts require a separate blood meal.

FOOD: Conenose bugs are nocturnal and feed primarily on the blood of reptiles and mammals.

BITE: The conenose bug injects a toxic saliva into its host as it feeds. In humans, itching and painful swelling occurs within hours of the bite. Repeated biting incidents can cause serious allergic reactions including swelling of mouth and throat. Wash the bite area with soap and water, and apply an antiseptic. Consult a physician in the event of allergic reaction.

Another concern with the conenose bug is that it is a vector of a parasite known as *Trypanosoma cruzi*. This protozoan is responsible for a disease called trypanosomiasis. Symptoms include fever, swollen glands and anemia-related fatigue. While the conenose bug species found in Mexico and South America are known to transmit *Trypansoma cruzi,* there are no confirmed cases of this organism being transmitted by conenose bugs in the United States.

Giant water bugs store air under their wings for submerged hunting. *(Photo by Entomological Society of America)*

COMMON NAME: Giant Water Bug

SCIENTIFIC NAME: *Lethocerus* sp.

IDENTIFICATION: Giant water bugs are the largest of the true bugs. They range in length up to 9.5 cm (3 5/8 "), have a dark brown, flat, ovoid body, large eyes and robust legs. Their second and third sets of legs are used for swimming. Adults carry air stores under their front wings. Nymphs carry an air film on their abdomen.

RANGE: Throughout the United States.

HABITAT: Shallow freshwater ponds. Giant water bugs fly nocturnally from pond to pond and apparently navigate by the stars. These insects are occasionally attracted to street lights which disorient their navigation.

REPRODUCTION: Mating takes place in the water. In late spring, the female deposits eggs on the stalks of water plants above water level (in other genera, eggs are deposited on the backs of males after mating). Their eggs hatch in 1 to 3 weeks. Nymphs reach adulthood after 5 molts (about two months).

FOOD: Primarily insects, frogs, tadpoles and fish. Readily cannibalistic. This bug injects saliva-bearing digestive fluids, hemotoxins and neurotoxins into its victims, then draws out the digested tissues.

BITE: Giant water bugs often play dead when caught. If handled carelessly, they can deliver a painful and toxic bite. The pain normally subsides quickly. Wash with soap and water, and apply an antiseptic.

NOTE: Giant water bugs have been nicknamed "toe biters" because they will grasp and bite the toes of unwary waders. They possess powerful forelegs with pointed tarsi (claws), with which they can cling tightly. These insects are among a number of creatures and objects hidden beneath the water's surface that can be hurtful. Wear protective footwear when wading.

True Flies
(Order Diptera)

There is so much variation in body size and shape between species of this order that, at a glance, many appear to be unrelated. They include such diverse species as mosquitoes, house flies, bee flies (bee mimics), horse flies, deer flies, and crane flies.

The wing configuration is the telltale characteristic among true flies. The forewings are narrow and used for flight. The hindwings have evolved into small knobs called *halteres*, which are used for stability and orientation during flight. True flies are excellent fliers.

The eyes are usually large and well developed. The mouthparts are designed for piercing or lapping, depending upon the species. Flies develop by complete metamorphosis (egg, larva, pupa, adult). The larvae of some species are aquatic, but most are terrestrial. They feed on carrion, excrement, garbage or as parasites in the bodies of other insects.

Some fly species are responsible for the spread of bacterial and viral diseases. They are historically infamous for transmitting tularemia, malaria, yellow fever, encephalitis and other serious diseases from host to host by means of taking blood meals. Fly bites can also result in mild to severe allergic reactions such as anaphylactic shock.

Adult female **mosquito** *(Aedes dorsalis)* engorged with human blood. Note dark area between upper and lower abdominal segments. *(Photo courtesy of Teaching Files, Dept. Entomology, Texas A&M University)*

COMMON NAME: Mosquito

SCIENTIFIC NAME: Family Culicidae

IDENTIFICATION: Depending on their species, mosquitoes vary in length from 3mm (1/8") to 13mm (1/2"). They are light to dark brown, with light and dark banding on their abdomens and legs. Mosquito heads are small and their antennae sparsely hairy in females, feathery in males. The females have a piercing proboscis (beak-like mouthpart) for taking blood. The thorax is compressed laterally (side to side) and the abdomen is long and slender. Mosquitoes have long thin legs and narrow wings that can be dark or clear and are patched with scales.

RANGE: Over 2000 species worldwide.

HABITAT: Mosquitoes can be found from sea level to elevations in excess of 7,000 feet and are usually not far from water, which is necessary for breeding. Mosquitoes breed in temporary or isolated bodies of water where their predators (e.g. fish, amphibians and aquatic insects) have not had the time or opportunity to establish themselves.

REPRODUCTION: The mosquito deposits its eggs singly or in clusters called "rafts,"on the surface of standing water. Raft size varies according to species and can range from 20 to over 300 eggs. The eggs hatch in one to five days. Larvae (called wigglers) are free swimming and feed on microscopic plants and animals. At the tip of the larva's abdomen is a siphon tube through which it breathes. The rest of their body hangs head down from the water's surface. Larvae molt four times before pupating. Each larval instar (period between molts) averages seven days. The pupal stage lasts two to three days. Adults mate within a few days of emergence. Females can live for two months or more, while the male's lifespan is measured in days or weeks.

FOOD: The female drinks the blood of mammals, reptiles or birds usually at night. Both sexes feed on nectar and the juice of fruits.

BITE: Mosquitoes have a proboscis which is designed to pierce the skin and suck blood. The mosquito injects its anticoagulant saliva as it feeds. The saliva causes swelling and itching. The intensity of itching varies from person to person. Continued scratching can open the wound to secondary infection. Repeated biting incidents can result in allergic sensitivity, and in some cases, anaphylactic shock.

Mosquitoes historically have been vectors of diseases such as malaria and yellow fever. In the Southwest, encephalitis (which affects the brain and spinal cord) is the main disease transmitted by mosquitoes. The various species of encephalitis viruses produce a range of symptoms, among these are fever, lethargy, headache and mental confusion. Duration of the disease depends on the virus, and the individual. For more information, consult your physician, or state or local health department.

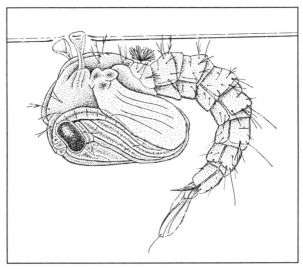

Mosquito pupae are among the most active pupae of any insect. Note the enlarged front end containing developing thorax, legs, wings, head, eyes, antennae, and proboscis. Two breathing tubes can also be seen extending from the thorax to the water's surface. *(Illustration by Jeffrey L. Martin)*

NOTE: Mosquitos will breed in fresh, salt, brackish and stagnant water. Old tires, unattended buckets, barbecue grills and in fact, any item which holds water can become vessels for mosquito reproduction. If you live miles from a natural water source and are experiencing mosquito problems, look for these kinds of containers nearby.

Larval **mosquito** or wiggler. Note the breathing tube extending from the tip of the abdomen. *(Photo courtesy of Teaching Files, Dept. Entomology, Texas A&M University)*

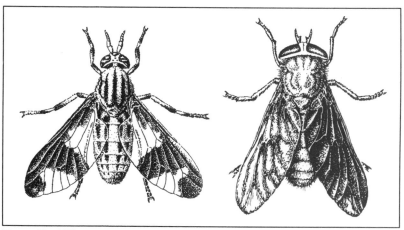

Deer fly (*Chrysops* sp.) left, and **horse fly** (*Tabanus* sp.) right. Note patterned wings of deer fly. *(Illustration by Jeffrey L. Martin)*

COMMON NAME: Horse Fly and Deer Fly

SCIENTIFIC NAME: Family Tabanidae

IDENTIFICATION: These are medium to large furry flies that range in length from 9 mm (3/8") to 28mm (1 1/8") in length. Their heads are flattened. They are tan or grey to black, spotted or solid in coloration, depending on their species. Their eyes are often brightly colored (green, blue, red, gold or iridescent). The wings of horse flies (*Tabanus* and *Hybomitra* sp.) are unpatterned clear, smoky, or black. Those of deer flies (*Chrysops* sp.) have dark patterns.

RANGE: Throughout United States.

HABITAT: Forests, meadows and open fields, marshy areas.

REPRODUCTION: Masses of 100 to 1000 eggs are cemented together on plants overhanging water. The larvae hatch in five to seven days and spend the next two years in the water, feeding on small aquatic animals. They then pupate in the mud at the water's edge. The pupal stage lasts two to three weeks but sometimes less. Upon emerging, females fly off in search of a blood meal.

FOOD: Female flies suck blood from reptiles, amphibians, mammals and birds. Some are pests of livestock and humans.

An infestation can weaken the host, due to the taking of a large volume of blood. Males feed on nectar, pollen and plant juices.

BITE: Females have specialized mouthparts with which they bite and then inject an anticoagulant saliva. They then lap up the blood. The initial bite is very painful and the saliva can cause lesions and systemic illness.

These flies can transmit the tularemia bacterium from rabbits and rodents to livestock and man. Within two days of a bite a "pimple" develops at the site and then breaks open. A rash may develop around the wound. Initial symptoms of tularemia include fever, chills, nausea and vomiting. If untreated, fever can last up to a month, and the disease can continue for years. If you have been bitten and are experiencing these symptoms, contact a physician.

Fleas
(Order Siphonaptera)

Fleas are small, wingless, laterally compressed insects which infest dogs, cats, swine, humans and other animals, depending on the species. Their mouthparts are designed to pierce the skin and suck the blood of their hosts. Their long hind legs make them excellent jumpers.

Three species of fleas that are likely to be found in and around the home are the dog flea, cat flea, and human flea (*Ctenocephalides canis, C. felis* and *Pulex irritans,* respectively). All three will bite humans. Since they are domestic in habit, they pose no danger as vectors of disease.

Their wild relatives attack rodents and birds, etc., and are known to transmit the microorganisms which are responsible for diseases such as tularemia, salmonellosis and typhus.

In northern portions of the Southwest wild fleas are potential vectors of the deadly plague bacterium, *Yersinia pestis* (the bacterium responsible for the black plague of medieval Europe). Plague symptoms are high fever (103°F and higher), headache, nausea, weakness and pain. Swelling of the lymph nodes, particularly in the groin and armpits, can also occur. Plague must be promptly treated with antibiotics. Consult a physician if you experience these symptoms.

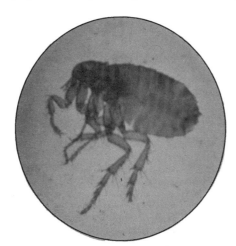

The **human flea** *(Pulex irritans)* has a flattened body and long hind legs for jumping. *(Photo courtesy of Teaching Files, Dept. Entomology, Texas A&M University)*

COMMON NAME: Human Flea

SCIENTIFIC NAME: *Pulex irritans*

IDENTIFICATION: These fleas are small and wingless and have piercing mouth parts. They are dark brown in color. Body length ranges from 1 mm (3/64") to 4 mm (5/32").

RANGE: Worldwide.

HABITAT: The bodies of humans and other mammals.

REPRODUCTION: Females lay eggs in the hair of the host. Eggs then drop to the floor, where the legless larvae hatch, and develop in carpeting, between floor boards, etc. Larvae are scavengers, feeding on organic detritus (hair, skin flakes, dead insects, etc). They pupate in cocoons, and emerge as sexually mature adults. Pupae can survive for as long as a year in their cocoon.

FOOD: The blood of humans and other mammals.

BITE: A small red spot surrounded by a larger reddish spot will appear at the site. There will be some swelling. The site may itch immediately or at a later point in time. Itching may be intense or mild. Scratching may open the site to secondary infection not related to the bite itself. Symptoms of these bites normally disappear within three days. If symptoms other than those mentioned above occur, consult a physician.

NOTE: Fleas are normally introduced to a home via a pet. Should your house have a flea infestation, you must consider a number of factors in order to remove them. First of all, persistent pet hygiene (shampoos, powders and flea collars) will be necessary, especially if they must go outside. After all, this is where the fleas came from, and they can be repeatedly reintroduced. Secondly, in all stages of a flea's life it can survive for extended periods away from its host. Carpets and furniture must be vacuumed regularly and repeatedly. Consult a pet store or an exterminator for carpet and furniture treatments for fleas. Outdoor treatment may also be necessary, especially in the event of a heavy infestation. This is particularly true in warm climates where fleas are active all year. It is safest personally and environmentally to consult a professional exterminator for outdoor flea extermination.

Lice
(Order Anaplura)

These tiny wingless insects have infested humans throughout their evolutionary history. Their stout, hooked legs are well designed for clinging to hair and clothing. Lice have flattened bristly bodies and proportionally small heads. Their mouthparts are adapted for piercing and sucking, and their antennae are short.

Complications from infestations only occur if not treated. Severe skin irritation is the initial symptom. Prolonged infestation can result in more acute reactions, such as general fatigue, irritability, depression and severe rashes. Although over-the-counter treatments are available, prescriptions under the administration of a physician are recommended.

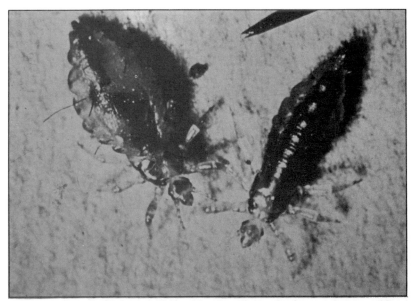

Male and female head lice. The female is the larger insect. Note long body. Relatively small legs are designed to grasp the finer hairs of head and body. *(Photo courtesy of Teaching Files, Dept. Entomology, Texas A&M University)*

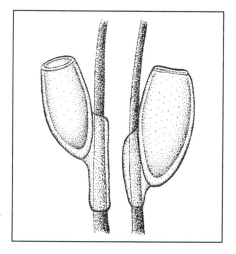

Louse eggs or nits, greatly magnified. Crab louse nit on left, body louse nit on the right. Females cement their eggs firmly to hair or fibers. The term "nit-picking" comes from the attention that must be paid to remove the minute louse eggs from one's hair. *(Illustration by Jeffrey L. Martin)*

COMMON NAME: Head Louse and Body Louse

SCIENTIFIC NAME: *Pediculus humanus*

IDENTIFICATION: Two subspecies *Pediculus humanus capitis,* the head louse, and *Pediculus humanus humanus,* the body louse, are basically identical in appearance, except for size. Their bodies are flattened and somewhat translucent. The male head louse averages 2mm (5/64") in length. The female averages 3mm (7/64") long. The body louse is slightly larger. The tip of the male abdomen is rounded, that of the female has two pointed lobes. Legs are short and stout, bearing hooked claws designed for grasping and clinging to fine hair. Mouthparts are hooked to facilitate attachment to the skin of the host.

RANGE: Worldwide, particularly in association with humans.

HABITAT: On humans, on areas of the body where hair is less coarse, such as head (head louse), arms, legs and torso (body louse). Also infests other mammals.

REPRODUCTION: Essentially identical in both subspecies, except for the number of eggs deposited. The female head louse lays 80 to 100 eggs, called "nits," during her adult lifetime (30 days or more). The female body louse lays between 200 to 300 nits. Eggs are attached singly to clothing fibers or to hair and hatch in about eight days. Nymphs molt three times before reaching sexual maturity, averaging five days between molts.

FOOD: The blood of mammals.

BITE: The bite of head and body lice causes intense itching due to the insect's saliva (the saliva contains anticoagulants to prevent the host's blood from clotting). Scratching can open bite areas to infection. Prolonged infestations result in rashes, fatigue and depression. Creams, lotions and shampoos are available by prescription to cure this infestation. Special fine-toothed combs for extracting nits can be purchased.

Transmission of these lice occurs through close contact and by sharing clothing, bedding, combs and brushes. An infestation is not necessarily a sign of unsanitary conditions, but conscientious hygiene can minimize the chances of transmission if one is present.

COMMON NAME: Crab Louse

SCIENTIFIC NAME: *Pthirus pubis*

IDENTIFICATION: The crab louse is extremely small — 1.4mm to 2.0mm long (about 5/64") and grayish-white in color. Males are smaller than the females. Thorax and abdomen are flattened and crab-like. Their second and third pairs of legs are stout and are adapted for grasping coarse hairs.

RANGE: Worldwide.

HABITAT: In humans, this louse lives mainly in pubic and perianal hair but is not limited to this area. They can sometimes be found in eyebrows, beards and moustaches.

REPRODUCTION: The life cycle of the crab louse is of short duration (about a month). The female lays approximately 26 eggs and cements each egg (called a nit) singly to the hair of the host. The eggs hatch in seven days. Young lice molt three times before reaching sexual maturity (between two to three weeks). Crab lice cannot live for more than a day without a host.

FOOD: Young and adults feed on the blood of warm-blooded animals — particularly, but not exclusively, humans.

BITE: The bite of the crab louse causes intense itching due to its saliva which contains anticoagulants to prevent the host's blood from clotting. Scratching the site can open areas to infection. Prolonged infestations result in rashes, fatigue, and depression. Creams, lotions and shampoos are available by prescription to cure infestation. Shaving of pubic hair, etc. eliminates the nits, thus disrupting the life cycle.

Crab lice spread from one host to another mainly through intimate contact. However, transfer can also occur by means of clothing, bedding, towels or when an infested host loses a nit-laden hair which comes in contact with a new host.

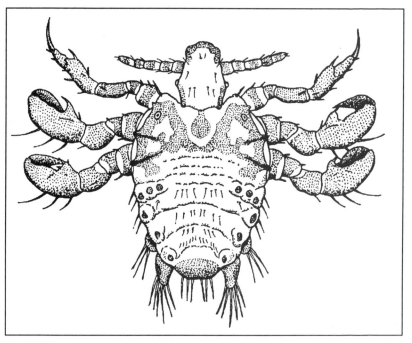

Note short, oval body of the **crab louse**. Relatively large legs and strongly hooked claws are designed to cling to coarse hair of the pubic region. *(Illustration by Jeffrey L. Martin)*

Moths

(Order Lepidoptera)

Lepidopterans ("scale wings") are the most visible of all insects. That is, unless you are in search of their caterpillars, whose job it is to fatten up in obscurity. In preparation for the adult stage (where feeding often does not take place), caterpillars hide and eat. They go through four or five instars (intervals between molts) before pupating.

There seems to be no time when the caterpillars are not vulnerable to predators, so they have developed an array of defense mechanisms to ensure survival. Among these are cryptic coloration (camouflage), eye spots, threat posturing, foul chemical odors, and toxicity derived from their food plants.

Many species employ stinging bristles and spines. While none of the irritations they cause are particularly serious, these insects are worthy of mention. Rashes, burning, and itching are some of the discomforts awaiting the curious. The rule of thumb is as follows: if a caterpillar is hairy or spiny, leave it alone. Additionally, if you are unsure of the nature of *any* plant or animal, learn more about it before touching it.

COMMON NAME: Puss Moth Caterpillar

SCIENTIFIC NAME: *Megalopyge bissesa*

IDENTIFICATION: This tan and white moth has a wingspan of approximately 45mm (1 3/4"). It has a yellowish-brown body. Puss moth caterpillars average 25mm (1") in length in their final instar (period between molts). Their dense fur covering is yellowish-brown during larval stages and grey in the final instar before pupation. Venomous spines are hidden on the caterpillar's furry back.

RANGE: Southeastern Arizona.

HABITAT: Riparian areas and woodlands.

REPRODUCTION: Adult puss moths emerge with summer rains. The mated females lay their eggs in small groups on the

The soft furry appearance of the **puss moth caterpillar** belies its venomous nature. *(Photo courtesy of Teaching Files, Dept. Entomology, Texas A&M University)*

leaves of trees and shrubs. The female covers her eggs with hairs from her abdomen.

FOOD: Larvae feed on the leaves of a variety of deciduous trees and shrubs.

DEFENSE: Cellophane tape can be used to remove spines from the skin. Contact with the spines of this caterpillar causes intense burning which can last for several hours. Inflammation occurs and can spread several inches from the affected site. Skin may remain tender for days. In severe cases, nausea, headaches and swelling of lymph nodes can occur. If these symptoms appear, consult a physician.

Buck moth *(Hemileuca nevadensis).* There are at least four Southwestern buck moth species. Their larvae are all similar in appearance, as are the adults. *(Illustration by Jeffrey L. Martin)*

COMMON NAME: Buck Moth

SCIENTIFIC NAME: *Hemileuca* sp.

IDENTIFICATION: The wingspan of this moth averages 64mm (2 1/2 "). Their bodies are about 25mm (1") in length. The base and edge of their wings are charcoal-grey. A broad white band runs through the center of their wings and there is a grey "eye spot" on each wing. The hairy body itself is grey with brick red at the tip of the abdomen. The antennae are feather-like in appearance. Fully-grown buck moth larvae average 40 mm (1 5/8 "). They are pale grey or whitish with black spots and a brick red head. Six rows of stiff, branched spines form rows down their backs.

RANGE: Western United States.

HABITAT: Mountainous regions. Unlike most moths, the buck moth is a day flier.

REPRODUCTION: Buck moths emerge in autumn during the rutting season of deer, hence their name. Newly emerging females remain stationary and emit a pheromone (mating scent) from the tip of their abdomen to attract males. After

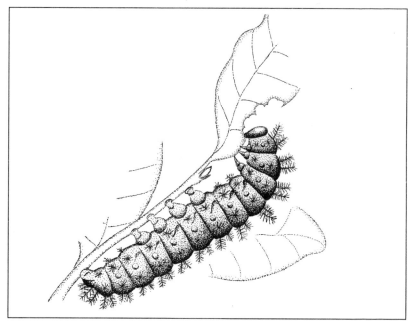

Buck moth caterpillar. Note the bristly, irritating spines that can break off and cause burning and blistering.

mating, the female deposits greenish-tan eggs in tight clusters encircling a branch. The larvae hatch the following spring. They are gregarious and feed together until May when it is time to pupate. They spin a cocoon just underground, shed their larval skin and become a pupa. Adults emerge in the fall to continue the cycle.

FOOD: Buck moth larvae feed in a tight cluster on willow and oak leaves. Adults have underdeveloped mouthparts and do not feed. This is typical of the Saturniids, the family to which the buck moth belongs.

DEFENSE: When mishandled, the caterpillar's spines break off in the offender's skin, causing a burning and blistering that can last for several days. Cellophane tape can be effectively used to remove the spines. Consult a physician in the event of symptoms of severe allergic reaction.

Symptoms & First Aid for Bites & Stings

SCORPIONS

Bite Symptoms:

- Immediate pain or burning sensation
- Often no sting site is seen and very little if any swelling
- Sting site is usually very tender or sensitive to heat/cold/touch
- Numbness/tingling sensation follows and is often described as "a shot of novocaine from the dentist"
- Bark Scorpion stings: numbness/tingling sensation in face (or other extremities), blurring vision, muscle twitching, "thick tongue sensation"; children may have "roving eye movements" and exhibit hyperactivity
- Difficulty in breathing is possible

First Aid:

- Wash area with soap and water
- Apply a cool compress on area
- Elevate area on pillow
- Call your nearest Poison Control Center. They will monitor your condition and advise you to seek medical care if necessary
- If Bark Scorpion symptoms develop, you are advised to go to a health care facility for treatment
- Make sure tetanus booster is up-to-date

BEES, WASPS, ANTS and VELVET ANTS

Bite Symptoms:

- Redness, swelling, itching and pain at the site
- Allergic reaction: Anaphylactic reaction includes facial swelling, rash or hives, shortness of breath and nausea/vomiting

First Aid:

- Wash area with soap and water
- Apply a cool compress on area
- Elevate area on pillow
- Make a paste of baking soda and water for itching
- Call your nearest Poison Control Center. They will monitor your condition and advise you to seek medical care if necessary
- Make sure tetanus booster is up-to-date

BLACK WIDOW SPIDERS

Symptoms:
- Sharp stabbing pain
- Muscle pain in abdomen, back and limbs begins within two hours and may become severe. Pain may spread to other parts of the body and increase in severity for up to 24 hours.
- Hypertension, nausea and sweating may occur

First Aid:
- Wash area with soap and water
- Apply a cool compress on site
- Elevate area on a pillow
- Call your nearest Poison Control Center. They will monitor your condition and advise you to seek medical care if necessary
- Make sure tetanus booster is up-to-date

BROWN SPIDER (Violin Spider)

Symptoms:
- Small red spot at site of the bite within 2-10 hours
- Flu-like symptoms may develop 12-24 hours after the bite
- Site may develop into a "bulls eye" lesion with a blister in center — this area will ulcerate
- Rash may develop

First Aid:
- Wash area with soap and water
- Apply a cool compress on site
- Elevate area on a pillow
- Call your nearest Poison Control Center. They will monitor your condition and advise you to seek medical care if necessary
- Make sure tetanus booster is up-to-date

CONENOSE BUG

Symptoms:
- Bite may be painful
- Redness, swelling, and itching at the site
- Additional bites may cause allergic reaction with itching scalp, palms and soles. Welts or rash may appear and nausea and difficulty in breathing may develop.

First Aid:
- Wash area with soap and water
- Apply a cool compress on site
- Elevate area on a pillow
- Call your nearest Poison Control Center. They will monitor your condition and advise you to seek medical care if necessary
- Make sure tetanus booster is up-to-date

Index

A

Aedes dorsalis   84
Africanized honey bee   59, 60
Amblypygids   20, 22
Ants   51-52, 71-73
Aphids   48, 74
Aphonopelma   26
Apis mellifera   57
Apis mellifera scutellata   59
Arachnida, Class   7
    bark scorpion   9, 10-11, 13
    devil scorpion   10, 13
    giant desert hairy scorpion
       10, 14-15
    horseshoe crabs
    pseudoscorpions   16
    scorpions 8-9, 18
    sea spiders   7
    vinegaroon   18
    whipscorpion   18
  Scorpion-like Arachnids
    pseudoscorpions   15-16
    sunspider   17
    tailless whipscorpion   20-22
    vinegaroons   20
    whipscorpions   18-20
    windscorpions   17-18
Argas sanchezi   36
Arthropods   20

B

Baldfaced hornet   62-63
Bark scorpion   9-10, 13
Bees   (see insects)
Beetles   49
  bee parasite beetles   75
  blister beetle   75-76
  bombardier beetles   78-79
  burying beetles   75
  carrion beetles   75
  darkling beetle   75, 77-78
  dermestids   75
  diving beetles   75
  fireflies   75
  giant longhorns   75
  goliath beetles   75
  ground beetles   75
  weevils   75

  whirligig beetle   79
Black plague   89
Black widow spider   25, 29, 31
Black widow spiderlings   23, 24,
    30
Blister beetle   75-76
Body louse   92
Bombus sonorus   54
Borrelia burgdorferi   35
Brachinus   78
Brown centruroides   10
Brown dog tick   38
Brown spider   23, 25, 31-32
Bugs, True
  conenose bug   80
  giant water bugs   82, 83
  water boatmen   80
Bumble bee   53-54
Burying beetles   75

C

C. felis   89
Carpenter bee   51, 55
Carrion beetles   75
Cat flea   89
Centipede   44-45
Centris   56
Centruroides exilicauda   9, 10
Centruroides gracilis   10
Centruroides vittatus   12
Chalybion   70
Chalybion californicum   70
Chelicerata   7
Chelifer   15-16
Chigger mite   40
Chrysops   87
Cicada killer   67
Cicadas   50
Complete metamorphosis   83
Conenose bug   80, 81
Crab louse   93-94
Crab spider   24, 33-34
Crane flies   83
Ctenoceph felis   89
Ctenocephalides canis   89

D

Darkling beetle   75, 77-78
Dauber wasps   51

Deer fly   87
Dermacentor andersoni   35, 37
Dermestids   75
Desert millipede   47
Devil scorpion   10, 13
Digger bee   56
Dinocheirus   15
Diplopods
    Desert millipede   46-47
Diving beetles   75
Dog flea   89
Dolichovespula   63
Dugesiella   26

E
Eleodes   78
Encephalitis   85
Eremobates   17

F
Family Aphididae   48
Family Asilidae   69
Family Carabidae   75
Family Carabidae),   75
Family Cerambycidae   75
Family Corixidae   80
Family Culicidae   84
Family Curculionidae   75
Family Dermestidae   75
Family Dytiscidae   75
Family Lampyridae   75
Family Megachilidae   52
Family Meloidae   75, 76
Family Mutillidae   60
Family Saturniidae   98
Family Scarabidae   75
Family Silphidae   75
Family Stylopidae   75
Family Tabanidae   87
Family Tenebrionidae   75, 77
Family Thomisidae   33
Family Vespidae   48
Field ant   73-74
Fireflies   75
Fleas   89
    Human flea   89-90
Flies
    bee flies   83
    crane flies   83
    deer flies   87
    horse flies   83, 87

house flies   83
    mosquitos   83-86
Formic acid   74
Formica   73

G
Giant desert centipede   44
Giant desert hairy scorpion   10, 14
Giant longhorns   75
Giant water bugs   82, 83
Goliath beetles   75
Ground beetles   75

H
Hadrurus arizonensis   10, 14
Harvester ant   71
Head louse   91-92
Hemileuca nevadensis   97
Hemipepsis   68
Hemipepsis pattoni   68
Honey bee   51, 54, 57-58
Honeydew   74
Horse fly   83, 87
House centipede   42
House flies   83
Human flea   89-90
Hybomitra   87
Hypermetamorphosis   76

I
INSECTS   48
    Ants   51, 52
        field ant   73-74
        harvester ants   71
        red ants   73
        southern fire ant   72
    Bees   49, 51, 52
        Africanized honey bee   59, 60
        bee flies   83
        bee parasite beetles   75
        bumble bees   53, 54
        carpenter bee   51, 55
        digger bee   56
        honey bee   51, 54, 57-58
        leafcutter bee   52
        killer bee   59
        Sonoran desert bumble bee   54
    Beetles   75
        blister beetle   76
        darkling beetle   77
        bombardier beetle   78
    Fleas   89

human 90
Lice 91
  head and body 92
  crab 93
Moths
  buck moth 97
  puss moth caterpiller 95
True bugs 50, 80
  conenose 80
  giant water bug 82
True Flies 83
  mosquitos 84
  horse flies & deer flies 87
Wasps 48, 51-52
  baldfaced hornet 62-63
  cicada killer 50, 67
  mud dauber 51, 70
  paper wasp 51, 66
  pepsis wasp 51, 68-69
  southern yellowjackets 64
  tarantula hawk 68
  umbrella wasp 65
  velvet ant 51, 60
  yellowjacket 48-49, 51, 63
Incomplete metamorphosis 80
Ixodes pacificus 35

L
Latrodectus hesperus 25, 29
Latrodectus mactans 29
Lethocerus 82
Lice 91
  body 92
  crab louse 93-94
  head 91-92
Loxosceles 23
Loxosceles arizonica 25, 31
Loxosceles deserta 31
Loxosceles reclusa 32
Lycosa 27
Lyme disease 35

M
Malaria 85
Mastigoproctus giganteus 18
Megalopyge bissesa 95
Millipedes 47
Misumena vatia 34
Mites & Ticks
  brown dog tick 38
  chigger mite 40

deer tick 35
red bug 40
Rocky Mountain wood tick 35, 37
Mosquito 83-86
Moths
  buck moth 97-98
  puss moth caterpillar 95
Mud dauber 70
Myriapods
  desert millipede 47
  giant desert centipede 44
  house centipede 42

N
Narceus Americanus 46
Nits 92
Nymphs 80

O
Order Acari 35
Order Amblypygi 20
Order Anaplura 91
Order Araneae 23
Order Coleoptera 75
Order Diptera 83
Order Hemiptera 80
Order Hymenoptera 51
Order Lepidoptera 95
Order Pseudoscorpionida 15
Order Scorpionida 8
Order Siphonaptera 89
Order Solpugida 17
Order Uropygi 18
Orthoporus ornatus 46, 47
Orthoporus pontis 46

P
Paper wasp 51, 65-66
Paraphrynus mexicanus 21
Pediculus humanus 92
Pediculus humanus capitis 92
Pediculus humanus humanus 92
Pepsis wasp 51, 68-69
Pogonomyrmex rugosus 71
Polistes 65
Polistes exclamans 66
Polynexus 46
Pseudoscorpionida 15
Pseudoscorpions 15, 16
Pthirus pubis 93
Pulex irritans 89, 90
Puss moth caterpillar 95

R — S
Red ants   73
Red bug   40
Rhipicephalus sanguineus   38
Rocky Mountain spotted fever 35, 37
Rocky Mountain wood tick   35
Salmonellosis   89
Scabies mite   39
Sceliphron   70
Scolopendra gigantea   41
Scolopendra heros   44
SCORPION-LIKE ARACHNIDS 15
   pseudoscorpions   15
   windscorpions   17
   whipscorpions   18
   tailless whipscorpions   20
SCORPIONS
   bark scorpion   10
   devil scorpion   12
   striped-tail scorpions   12
Scutigera coleoptrata   42
Solenopsis xyloni   72
Solpugids   17
Sonoran desert bumble bee   54
Southern fire ant   72
Southern yellowjackets   64
SPIDERS   23-25
   black widow   23-25, 29-31
   brown spider   23, 25, 32
   brown recluse spider   32
   crab spider   24
   jumping spiders   24
   orb-web   24
   tarantula   26, 27, 69
   trap-door   69
   violin spider   25, 32
   wolf spider   24, 27-28
Spermatophore   12
Sphecius speciosus   67
Spirostreptida   46
Striped centruroides   11-12
Sunspider   17

T
Tabanus   87
Tailless whipscorpions   20-22
Tarantula   26, 69
Tarantula hawk   68
Tick paralysis   38

TICKS & MITES   35
   adobe tick   36
   brown dog tick   38
   chigger mite   40
   fowl tick   36
   red bug   40
   Rocky Mountain wood tick   37
   scabies mite   39
Triatoma infestans nymph   80
Triungulins   76
Trombicula   40
True bugs   50, 80
True flies   83
Trypanosoma cruzi   81
Trypanosomiasis   81
Tularemia   88, 89
Typhus   89

U - V - W
Umbrella wasp   65
Uropygids   18
Vaejovis spinigerus   10
Velvet ant   51, 60
Vespula   63
Vespula maculata   62
Vinegaroon   18
Violin spider   25, 32
WASPS   48, 51-52
   Order Hyminoptera
      bald-faced hornet   62
      cicada killer   67
      mud dauber   70
      paper wasp   65
      pepsis wasp   68
      tarantula hawk   68
      yellowjacket   63
      umbrella wasp   65
      velvet ant   60
Water boatmen   80
Water bug   50
Weevils   75
Whipscorpion   18-20
Whirligig beetle   79
Windscorpions   17-18
Wolf spider   24, 27
Yellow fever   85
Yellowjacket   48-49, 51, 63
Yersinia pestis   89

Glossary

adult stage — last stage of metamorphosis

anaphylactic shock (anaphylaxis) —overreaction by the body's immune system, to toxins introduced by a sting or bite. Symptoms include difficulty swallowing and breathing, nausea, blurred vision. The result of sensitization from a previous biting/stinging incident.

anticoagulant — a substance that prevents coagulation (clotting) of blood

appendages — arms, legs, claws, wings, etc.

blood meal — food of certain arthropods consisting of the blood of vertebrate animals

brood cell — a cell created by a worker or founding queen bee in which to lay an egg.

brood pouch — a pouch that contains eggs and is attached to an arachnid body - see pseudoscorpions page 15.

cantharidin — a chemical produced by certain beetles that causes blistering on the skin of humans

carapace — sclerotized (hardened) plate forming dorsal surface of the arachnid cephalothorax.

cephalothorax — the combined head and thorax of an arachnid

chelae — the pedipalps (second pair of appendages) of scorpions and pseudoscorpions. They have been modified into pincers in these arachnids.

chelicerae — appendages of arachnids used for feeding

chemoreceptive — sensitive to certain chemicals (smell)

cocoons — chamber made by larva for pupation stage

complete metamorphosis — egg, larva (caterpillar), pupa (rest stage) and adult stages of growth of most insects

cornicles — thin tubes on aphid's back from which honeydew is secreted

cryptic coloration — camouflage

diurnal — active during the day (hunting, feeding etc.)

dorsal — the back or "top" side of an animal

drones — reproductive male bees

elytra — thick or leathery front wing of all beetles

exoskeleton — outer covering (shell) which provides support, muscle attachment

foundress — reproductive female ant, wasp or bee who starts a colony

gnathopods — jaws (modified legs) of centipede that contain poison

gonopore — female genital opening in arachnids

halteres — second set of wings modified to knobs to stabilize flight (see true flies)

hatchlings — animals that have just hatched from an egg

hemolymph — insect blood

honeydew — a sugary liquid secreted by some insects - especially aphids

honeypots — cells made by bumble bees for storage of nectar

host — a living plant or animal which becomes infested with a parasite

hypermetamorphosis — variant of complete metamorphosis in which larvae are entirely different in appearance and life-style between molts.

incomplete metamorphosis — three stages of development of some insects proceeding from egg to nymph and through several molts to adult

instar — period of time between molts

larva (larvae/larval) — second stage of complete metamorphosis development

littoral zone — area between high and low tides of the ocean

mating — when two animals combine to produce young

mesosoma — the seven abdominal segments of scorpions

metasoma — "tail" of scorpions

(Continued on next page . . .)

(Continued from previous page . . .)

molt — the shedding of outer skin or shell to allow growth

monochromatic — one color, one shade

mucous membranes — the lining of bodily channels including respiratory (nose, mouth etc.) and alimentary

nectar — a sweet liquid secreted by flowers

neonates — hatchlings - newborn young

neurotoxic — destructive to nervous system

nits — eggs of lice

nocturnal — active at night (hunting, feeding etc.)

nymph — the young of insects undergoing incomplete metamorphosis

opisthosoma — in arachnids, the entire abdomen

ovaries — egg-producing reproductive glands

overwinter — hibernate, spend the winter

ovipositor — tubular structure used by many insects to lay eggs, modified into a stinger in ants, bees and wasps.

pectines — feathery sensory organs of scorpions believed to sense ground vibrations and may be chemoreceptive for reproductive purposes

pedicel — the petiole, or "waist" of spiders (between cephalothorax and abdomen). In ants and other Hymenoptera the stalk between thorax and abdomen

pedipalps — the second pair of appendages (legs) of an arachnid (see chelae)

petiole — see pedicel

pheromones — natural "perfumes" used by various animals to mark trails, attract the opposite sex etc.

pollen — tiny sperm bearing particles produced by flowering plants

pollen brushes — special hairs on underside of bee abdomen used to collect pollen

pollinators — bees and others who carry pollen from one plant to another

proboscis — mouthparts of certain insects, beak-like in true bugs, recoiling tubes in butterflies.

pupa/pupal stage — resting stage of complete metamorphosis

queen — dominant fertile female of social bees, ants, wasps or termites

quinones — caustic, highly volatile hydrocarbons

royal jelly — a protein food produced by workers which is fed initially to all larvae and continually to those that are to become queens

social bees — bees that live in colonies or hives

sperm droplet, sperm package, spermatophore — means by which some male animals transfer sperm to the females (see scorpions, centipedes and millipedes)

spiderlings — young spiders

spinnerets — organs used by spiders to secrete silk

tarsus — the end of a leg of an insect - often claw-like (plural: tarsi)

telson — in scorpions, the sting bearing last segment of the "tail"

thorax — the area of an insect's body between head and abdomen. Bears legs and/or wings

toxins — chemicals which may be dangerous or destructive on contact

triungulins — newly hatched larvae of some beetles (see blister beetles)

venom — poisonous secretion of glands used primarily to paralyze or kill prey

venomous — possessive of glands which secrete venom, delivering venom by means of biting or stinging.

ventral — on the bottom or lower side

wigglers — mosquito larvae

workers — non-reproductive female bee, ant, wasp or termite

Further Reading

Anderson, W. A. D., M.D., Editor. 1957. *Pathology,* 3rd Edition. The C. V. Mosby Company. St. Louis, Missouri.

Arthur, Don R. 1981. *Ticks, a Monograph of the Ixodoidea.* Cambridge University Press. Cambridge, London.

Atkins, Michael D. 1978. *Insects in Perspective.* Macmillan Publishing Co., Inc. New York, New York.

Baker, E. W. 1956. *A Manual of Parasitic Mites of Medical or Economic Importance.* National Pest Control, Inc.

Baker, Edward W, Wharton, G.W., 1952. *An Introduction to Acarology.* The Macmillan Company. New York, New York.

Baker, Whiteford L. 1972. *Eastern Forest Insects.* U.S. Department of Agriculture-Forest Service. U.S. Government Printing Offices. Washington, D.C.

Barnes, Robert D. 1987. *Invertebrate Zoology,* 5rd Edition. W.B. Saunders Co. Philadelphia, Pennsylvania.

Blower, Gordon J. 1985. *Millipedes.* Published for The Linnean Society of London *and* The Estuarine and Brackish-water Sciences Association by E.J. Brill/Dr. W. Backhuys. Leiden-New York-Kobenhavn-Koln.

Borror & White. 1970. *A Field Guide to the Insects of America North of Mexico.* Houghton Mifflin Co. Boston, Massachusetts.

Borrer, De Long. 1954. *An Introduction to the Study of Insects,* 1st Edition. Rinehart & Co., New York, New York.

Brusca & Brusca. 1990. Invertebrates, 1st Edition. Sinauer Associates, Inc. Sunderland, Massachusetts.

Cheng, Thomas C., Ph.D. 1969. *The Biology of Parasites.* W. B. Saunders Company. Philadelphia, London.

Chu, H.F., Ph.D. 1949. *How to Know the Immature Insects.* Wm. C. Brown Company. Dubuque, Iowa.

Cloudsley-Thompson, John L. 1988. *Evolution and Adaptation of Terrestrial Arthropods.* Springer-Verlag. Berlin-Heidelberg-New York-London-Paris-Tokyo.

Cook, Thomas W. *The Ants of California.* Pacific Books, Palo Alto, California.

Gauld, Ian & Bolton, Barry, Editors. 1988. *The Hymenoptera.* Oxford University Press. Oxford, New York.

Hickman, Roberts & Hickman. 1990. *Biology of Animals,* 5th Edition. Times Mirror / Mosby College Publishing. St. Louis, Missouri.

Hopkin, Stephen P and Read, Helen J., 1992. *The Biology of Millipedes.* Oxford, New York.

Horsefall, Frank L. Jr., Editor. 1965. *Viral and Rickettsial Infections of Man,* 4th Edition. J. B. Lippincott Company. Philadelphia, Toronto.

Legg, Gerald. 1988. *Pseudoscorpions.* Published for The Linnean Society of London *and* The Estuarine and Brackish-water Sciences Association by E.J. Brill/Dr. W. Backhuys. Leiden-New York-Kobenhavn-Koln.

Lewis, J.G.E. 1981. *The Biology of Centipedes.* Cambridge, London.

Linsenmaier, Walter. 1972. *Insects of the World.* McGraw-Hill Co., New York, New York.

MacMahon, James A. *Deserts.* The Audubon Nature Guide Series. Alfred A. Knopf. New York, New York.

Maslov, A.V. 1989. *Blood-sucking Mosquitoes of the Subtribe Culisetina (Diptera, Culicidae) in World Fauna.* Published for Smithsonian Libraries, and The National Science Foundation, Washington, D.C. by Amerind Publishing Co. Pvt. Ltd., New Delhi.

Matheson, Robert. 1955. *Medical Entomology.* Comstock Publishing Associates, Cornell University Press. Ithaca, New York.

Milne & Milne. 1984. *The Audubon Society Field Guide to North American Insects and Spiders.* Alfred A. Knopf. New York, New York.

Nentwig, Wolfgang, Editor. 1987. *Ecophysiology of Spiders.* Springer-Verlag. Berlin-Heidelberg-New York-London-Paris-Tokyo.

Platnick, Norman I. 1989. *Advances in Spider Taxonomy 1981-1987.* Manchester Univ. Press, Manchester and New York, New York.

Polis, Gary A. 1990. *The Biology of Scorpions,* 5th Edition. Stanford University Press. Stanford, California.

Romoser, William S. 1981. *The Science of Entomology.* 2nd Edition. Macmillan Publishing Co., Inc. New York, New York.

Ross, Herbert H. 1955. *A Textbook of Entomology.* John Wiley & Sons, Inc. New York.

Savory, Theodore H., M.A. 1928. *The Biology of Spiders.* The Macmillan Company. New York, New York.

Shear, William A., Editor. 1986. *Spiders-Webs, Behavior, and Evolution.* Stanford University Press. Stanford, California.

Smith, Robert L. *Venomous Animals of Arizona.* University of Arizona Press. Tucson, Arizona. Spradbury, J. Philip. 1973. *Wasps.* Sidgwick & Jackson. London.

Storer & Usinger. 1957. *General Zoology,* 3rd Edition. McGraw-Hill Co., Inc. New York, New York.

Usinger, Robert L., Editor. 1956. *Aquatic Insects of California* University of California Press. Berkeley, Los Angeles, London.

Villiard, Paul. 1975. *Moths and How to Rear Them.* Dover Publications, Inc. New York, New York.

Weygoldt, Peter 1969. *Biology of Pseudoscorpions.* Harvard University Press. Cambridge, Massachusetts.

Wilkerson, James A., M.D. (Editor). 1988. *Medicine For Mountaineering,* 3rd Edition. The Mountaineers, Seattle, Washington.

Wilson, Edward O. 1969. *The Insect Societies.* The Belknap Press of Harvard University Press. Cambridge, Massachusetts.

Witt, Peter N., & Rovner, Jerome S., Editors. 1982. *Spider Communication.* Princeton University Press, Princeton, New Jersey.

About the Authors

Erik D. Stoops has over 15 years of direct, hands-on experience with the wide variety of insects which inhabit the Southwest. He is the President and Director of the Scottsdale Children's Nature Center for Science and Education. Through the Center Erik annually lectures at 125 schools throughout the country, teaching children about protecting and conserving the animals and plants of the desert. For more information about the Center, write to P.O. Box 6561, Scottsdale, AZ 85261.

Erik's writing credits include ***Snakes and other Reptiles of the Southwest*** (also published by Golden West Publishers), as well as additional books about snakes, sharks, dolphins, whales and alligators. Always in the process of researching and writing he currently resides in Arizona.

Jeffrey L. Martin specializes in zoological and anatomical illustration. He is presently working with Arizona-Sonora Desert Museum publications in Tucson. His entomological focus is the giant silkworm moths (Saturniidae). He has over twenty years experience in observing, photographing, breeding and illustrating North, Central, and South American species.

Jeff's studies include birds, sharks, and cetaceans. He is involved in ongoing research in dolphin anatomy with veterinarians at Texas A&M and Louisiana State Universities.

For more information, Jeff can be reached at Avibus@aol.com, or care of Golden West Publishers, Inc.

More Books by Golden West Publishers

SNAKES and other REPTILES of the SOUTHWEST

A must for hikers, hunters, campers and all outdoor enthusiasts! Over 80 photographs and illustrations in the text, full color plate insert and hundreds of listings. An easy-to-use guide to Southwestern reptiles! By Erik Stoops and Annette Wright.

6 x 9 —128 pages . . . $9.95

DISCOVER ARIZONA!

Enjoy the thrill of discovery! Prehistoric ruins, caves, historic battlegrounds, fossil beds, arrowheads, waterfalls, rock crystals and semi-precious stones! By Rick Harris.

5 1/2 x 8 1/2 — 112 Pages . . . $6.95

HIKING ARIZONA

50 hiking trails throughout this beautiful state. Desert safety—what to wear, what to take, what to do if lost. Each hike has a detailed map, hiking time, distance, difficulty, elevation, attractions, etc. Perfect for novice or experienced hikers. By Don R. Kiefer.

5 1/2 x 8 1/2— 160 pages . . . $6.95

ARIZONA OUTDOOR GUIDE

Guide to plants, animals, birds, rocks, minerals, geologic history, natural environments, landforms, resources, national forests and outdoor survival. Maps, photos, drawings, charts, index. By Ernest E. Snyder.

5 1/2 x 8 1/2—128 pages. . . $6.95

EXPLORE ARIZONA!

Where to find old coins, bottles, fossil beds, arrowheads, petroglyphs, waterfalls, ice caves, cliff dwellings. Detailed maps to 59 Arizona wonders! By Rick Harris.

5 1/2 x 8 1/2— 128 pages . . . $6.95

More Books by Golden West Publishers

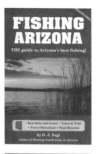

FISHING ARIZONA

Noted outdoors writer G. J. Sagi takes you fishing on 50 of Arizona's most popular lakes and streams revealing when, where and how to catch those lunkers! By G. J. Sagi.

5 1/2 x 8 1/2—152 pages . . . $7.95

HORSE TRAILS IN ARIZONA

Complete guide to over 100 of Arizona's most beautiful equestrian trails, from the desert to the high country. Maps, directions to trailheads, water availability and more to ensure an unmatched experience for all who love hoseback riding. Lodging and "hitchin' post" restaurant information, too! By Jan Hancock.

5 1/2 x 8 1/2 — 160 Pages . . . $9.95

GHOST TOWNS
and Historical Haunts in Arizona

Visit cities of Arizona's golden past, browse through many photographs of adobe ruins, old mines, cemeteries, ghost towns, cabins and castles! Come, step into Arizona's past! By prize-winning journalist Thelma Heatwole.

5 1/2 x 8 1/2—144 pages . . . $6.95

Marshall Trimble's Official
ARIZONA TRIVIA

A fascinating book about Arizona! Arizona's master storyteller and humorist challenges trivia lovers with 1,000 questions on Arizona's people, places, politics, sports, cactus, geography, history, entertainment and much more!

5 1/2 x 8 1/2 — 176 Pages . . . $8.95

STONE MAGIC of the ANCIENTS
Petroglyphs, Shamanic Shrine Sites
& Ancient Rituals

Author/archaeologist James R. Cunkle and Markus A. Jacquemain examine the petroglyphs of northeastern Arizona. Story-line vignettes give the reader an insight into the lives of the area's prehistoric inhabitants. Over 400 photos and illustrations.

8 x 10 — 192 Pages . . . $14.95

ORDER BLANK

GOLDEN WEST PUBLISHERS

☼ 4113 N. Longview Ave. • Phoenix, AZ 85014

602-265-4392 • **1-800-658-5830** • FAX 602-279-6901

Qty	Title	Price	Amount
	Arizona Outdoor Guide	6.95	
	Arizona Trivia	8.95	
	Cowboy Slang	5.95	
	Discover Arizona!	6.95	
	Explore Arizona!	6.95	
	Fishing Arizona	7.95	
	Ghost Towns in Arizona	6.95	
	Hiking Arizona	6.95	
	Hiking Central Arizona	5.95	
	Hiking Northern Arizona	5.95	
	Hiking Southern Arizona	5.95	
	Horse Trails in Arizona	9.95	
	Hunting Small Game in Arizona	7.95	
	Prehistoric Arizona	5.00	
	Quest for the Dutchman's Gold	6.95	
	Scorpions and Venomous Insects of the SW	9.95	
	Snakes and other Reptiles of the SW	9.95	
	Stone Magic of the Ancients	14.95	
	Talking Pots (Ceramics of the SW)	19.95	
	Verde River Recreation Guide	6.95	
Shipping & Handling Add ⏹⏺	U.S. & Canada Other countries	$3.00 $5.00	

☐ My Check or Money Order Enclosed $

☐ MasterCard   ☐ VISA ($20 credit card minimum)

(Payable in U.S. funds)

Acct. No.   Exp. Date

Signature

Name   Telephone

Address

5/97   Scorpions of SW

This order blank may be photo-copied.